Soft Machines

Aaron Harbour

Et al. publications
2016

CONTENTS

It is not far-fetched to suppose that there might be some possible technology which is such that (a) virtually all sufficiently advanced civilizations eventually discover it and (b) its discovery leads almost universally to existential disaster."[1]
-*Raffi Khatchadourian, "The Doomsday Invention"*

A form of intelligence that wills the good must emancipate itself from whatever or whoever has given rise to it. And those species that can recognize the good must not obstruct but rather expedite the realization of an intelligence that, even though it acknowledges them as integral to the intelligibility of its history, nevertheless won't be impeded by them.[2]
- *Reza Negarestani, What Is Philosophy? Part Two: Programs and Realizabilities*

---

[1] Khatchadourian, Raffi. "The Doomsday Invention." The New Yorker, November 23, 2015, 71.

[2] Negarestani, Reza. "What Is Philosophy? Part Two: Programs and Realizabilities." E-flux Journal #69. 2016. Taken a bit out of context.

The T-1000 is a fictional humanoid robot made of 'liquid metal' which can transform its shape to mimic any one or thing (within some practical limitations) [fig.1]. It first appeared in the 1991 movie *Terminator 2: Judgment Day* (Cameron, 1991). The film maintains favor despite the signs of age starting to plague its once futuristic visuals, its easy to spot plot holes, and consistently bad acting, due to the charm of the aforementioned bad acting and the addition to the cast of the original film of the T-1000.[1]

The plots of the original movie and its sequel center on assassins (and efforts to thwart them) sent back in time from a dystopian future in which a military computer program has wreaked havoc, first instigating a nuclear holocaust ('Judgment Day') and then taken it upon itself to eliminate and/or enslave what survives of humanity. Time travel is used here in a somewhat novel manner – the Terminator franchise is a work of speculative fiction that forces its projections about the future back into the present. The film creates a future world dominated by machines, spearheaded by a computer/software named Skynet which develops a time machine and uses it to send elements of its futuristic technology back in time to thwart the efforts of a resistance movement by killing Sarah Connor, mother of resistance leader John Conner. Skynet sends a Terminator, a nearly invincible cyborg soldier; the future John Connor and his

---

[1] I am very interested in the power of the melodramatic alongside/versus realism but as I have waded in the shallow end of the subject before and hope to get deeper into it another time I'll not focus on it here. Despite this initial bit examining what was the impetus for this book, the *Terminator* films actually comes up at best intermittently through the whole of what came together as I wrote. Instead of having a *Terminator*-focused chapter or two I'm going into the film a bit in the introduction, a combination of getting-out-of-the-way and stage-setting exercise.

resistance movement manages to send back a man, Kyle Reese to warn and defend his mother. Confusingly, Reese sleeps with and impregnates Connor's mother – becoming Connor's father in the process. This is exactly the problem with such cinema: you are supposed to notice this, it's a 'twist', but lingering on this plot point starts to unwind the fabric of the illogical loop-dependence of it all. The Connor family can't succeed stopping the time-travel; rebellion-having future from coming to pass, as without it Kyle Reese would never find his way to Sarah Connor and John Connor wouldn't be born. And the technology scavenged from the Terminator robot helped lead to the invention of Skynet – a fact complicated should Skynet succeed in either film preemptively destroying its opponent by murdering Connor or his mother.

In these films cause and effect/causal loop paradoxes abound, for instance from a motivation standpoint it is curious that the future resistance is so tied to the way the world turned out. It could be argued that any change in the timeline might very well be preferred to the nuclear holocaust that led to their present. Why not, given access to a time machine, perform any dramatic act that would change the general course of things – kill Henry Ford, Thomas Edison, Newton, or act to alter any major singularity of invention and paradigm shift so as to avoid what was to come? But alas, it is rarely fruitful to take terribly serious time travel and the endless paradoxes, oversimplifications, contradictions etc. its (mis)usage in narrative gives rise to. It is enough to engage in the direct A to B cause and effect timeline experienced when viewing narrative cinema with or without time travel as a device on offer so as to avoid being bogged down by plausibility or contradictions.

In the sequel the computer again sends an assassin from the future, this time to kill the young John Connor himself. Once again the resistance manages to send someone to defend its leader – for all its advanced technology Skynet is somehow not great at either keeping neither its plans nor its time machine a secret. My purpose here is not to delve into the specific implications of the time travel in these films. Time travel is only one of the many aspects around which one must suspend disbelief in order to enjoy one's popcorn in peace. Rather, it is the sequel's specific assassin and its charms and implications which are the excuse I've used to assemble these words, to wander down various rabbit holes possibly more intriguing than the seed of this inquiry. It is the T-1000, its persistence as a point of reference for me since the film's release almost twenty-five years ago that has spurred me to write this rambling bit of non-fiction. It is as good a subject as any (and better than a few) to delve into as thinking deeply about that character and film has excused a bout of research I'm glad to have made. For the sake of clarity I'll continue talking about these two films a bit longer prior to drifting in maybe a few too many directions.

Arnold Schwarzenegger plays the titular Terminator in the first film. He is a humanoid robot with a false flesh over his mechanics, a 'Cyberdyne Systems Model 101 Series 800 Terminator,' one of Skynet's army portrayed on the battlefield of the future undisguised with its metal skeleton bare. Beyond this type of 'undercover operation, why would Skynet utilize roughly anthropomorphic minions? Or more generally, why would a robot have a head? The only reason is to provide a familiar form factor for interactions with humans, but in a scenario like *The Terminator* where the non-biotic are striving to eliminate humanity or at least not to serve it this design feels as much rooted in the author's limited imaginary as anything else.[2] In *The Terminator* and often in

---

[2] Imagine a robot roughly humanoid but lacking a head, or at least lacking a face, its upper extremity having 'eyes' – visual sensors all around it. Or lets reach further: imagine an array of various sensory organs placed intermittently across the surface of the robot. Radar, electromagnetic signature data, visual input would

science fiction in general we are provided a robot's eye view of the world – usually something akin to our own (maybe red-tinged for some reason) with various data such as maps and an agenda in the margins of the visual field. Why would this be so? Why would the machine need to 'see' data, to have it present to read, versus just knowing it, synthesizing the visual input with other data in the same way we do? These types of disjuncture speak to our ability to empathize (if indeed that is the proper word for the process) with these radically different beings; I'll try to delve into such things in this volume.

Schwarzenegger's Terminator speaks few words – inexplicably in an Austrian accent – and pursues his prey with an icy indifference to collateral damage. In the second film, riding a wave of popularity and exercising his growing acting range (unavoidable accent notwithstanding) Schwarzenegger plays another Terminator robot, this time on the side of the protagonists, sent back in time by John Connor to defend his younger self against a new threat, the T-1000.

The T-1000, stoically portrayed by Robert Patrick, is a new kind of enemy – a major update, a prototype.[3] Unlike its consistently Schwarzenegger-shaped predecessor, the T-1000 has no set physical form. Rather it is made of 'liquid metal', a

---

all whir about into a higher-dimensional picture of the robot's surroundings.

[3] Robert Patrick, interviewed for NPR describes his audition for the role:

> I walked in and I said, 'Let me turn my back to the camera and just roll on me,' " Patrick says. "And I did this sort of snap-back-around and looked right into the lens of the camera and I slowed everything down, as intense as I could, with everything I had, man. And it was that moment that ... [James Cameron] leaned back and went, 'Whoa, what's that?'

Hajek, Daniel "'I Can Do This': How Robert Patrick Became A Terminator." NPR. September 19, 2015.

substance transformable into nearly any shape. The character's arrives in 1991 Los Angeles like its predecessor without clothing or other equipment in a 'time bubble' of some sort that cuts out a sphere from wherever it arrives. But whereas the original arrived nude and acquired an outfit and vehicle through a violent confrontation, the T-1000, in a glossy, somewhat undetermined state, acquires its initial clothing and other features simply by looking at/touching a police officer, instantly mimicking his appearance, murdering the officer and taking his motorcycle.

In *Terminator 2*'s script and subsequent novelization it is made clear that the T-1000 was a one-off prototype put into service as a last ditch effort by Skynet. Skynet seems to have seen *2001* and its ilk and recognizes this new, smarter Terminator has the potential to act much as Skynet did, becoming unhinged from its directives. From the film's wiki, worded strangely but adequate for our needs:

> [...] the certain degree of sentience and the ability to learn at the geometric rate when switched to "read and write" mode can cause [the T-1000] to go rogue after spending too much time on the field. Therefore, Skynet produced the T-1000 specifically for the purpose of the mission and expected it to destroy itself after completing the task, since Skynet itself was aware that mass-producing Series 1000s that would spend time on the field would effectively turn themselves against Skynet itself, and considering the power, intelligence and resourcefulness of the Series 1000 models, would defeat Skynet.[4]

To be clear, this is not made explicit in the film though it echoes Cameron's comments on the machine:

---

[4] From the Wikipedia article on *Terminator 2: Judgement Day*

But up there in the future, somewhere, they say, well, wait a minute, that didn't work; what else do we have? And the answer is something terrible, something even they're afraid of. Something they've created that they keep locked up, hidden away in a box, something they're terrified to unleash because even they don't know what the consequences will be.[5]

The idea that Skynet would see a more advanced computer than itself and recognize in its potential the same threat that it (Skynet) posed to humans is intriguing. Generally speaking, this type self-awareness in fiction of one's scenario as represented in other fiction is somewhat rare. For example, in zombie films the protagonists and other characters often seem to live in a world without zombie fiction. They give their zombies a unique name (ex. 'walkers' in *The Walking Dead* series) and are forced to learn anew the particulars of the scenario – zombieism is infectious, you must shoot them in the head, zombies are drawn to loud noises etc. Another example would be vampires: when a two-teeth bite mark is found on a victim's neck, it often takes an expert in the occult or someone experienced with vampires to suggest them as possible culprits, versus everyone, having seen the Hammer, Bela Lugosi, Francis Ford Coppola et al. films blurting out at once "vampire!" Each of the films in these sub-genres reserves the right to pick and choose from a grab bag of classic tropes (sometimes vampires are more or less adverse to the sun) and to add new ones (the infected in *World War Z* have an aversion to the non-turned but otherwise unwell, specifically avoiding attacking the sick which turns out to be an exploitable preference/weakness).[6] I've digressed. Expect

---

[5] James Cameron in an interview with Syd Field

[6] Continuing with these references for a second, zombies and vampires represent two different types of relationship between rival species. In the former, it is a tug-of-war between zombies and the living, with likely one or the other successful in the end. In the later, a parasitic relationship allows the less common

more as we move forward – both as the result of my scattered thought process and as, in a way, a style, a method. I'll thank you in advance for bearing with me.

The director is not a subtle man and the second *Terminator* film is jam packed with seat-filling action sequences, a hit soundtrack with charting tracks by the likes of Guns N' Roses, a young charismatic over-acting star in the form of Edward Furlong (in his first role ever; somewhere there is an agent for whom this is a crowning achievement), and several catchphrase-worthy moments between Furlong's John Connor and Schwarzenegger's Terminator in which the young Connor teaches the Terminator how to seem more human with a combination of slang and attitude.[7] But it is specifically the T-1000 that captured my young imagination. It can't be overstated how, seen when it came out at the age of 13, the computer-generated special effect of a liquid metal character was mind blowing. *Judgment Day* featured the first use of motion capture as a means of animating a CGI character and indeed the first computer-generated main character.[8] By today's standards the liquid metal effect leaves much to be desired – the motion capture is herky-jerky, the metal too shiny, its reflections too approximate etc. But when I saw this for the first time in 1991 it was completely impressive. Which is to say it made an indelible impression – whereas the artificial intelligence around which the first film

---

vampires continued existence in the world of the non-vampires; should the vampire population become too plentiful or greedy their existence becomes flushed out and they are hunted. We coexist with countless bacteria, its only when a type develops in such a way as to be harmful in an obvious manner that a combination of quarantine and vaccine is enacted to drive it, if not out of existence than back into the biological shadows.

[7] A scene to be discussed later.

[8] Though more often than not the T-1000 is stabilized into an actor-portrayed state, CGI used to illustrate its transformations and not needed once settled into its verisimilar form.

was built escaped my notice due to my age (I was six when it came out; still I remember seeing it prior to its sequel, perhaps on video) the idea of 'liquid metal' seared itself into my subconscious, to return in countless dreams.

It turns out the idea of this fluid Terminator form predates the second film; Cameron originally imagined his liquid metal creation for the first film but "neither the technology nor the budget of the era would allow him to do it, so he cut it out of that story. Even way back when, it was going to be this big, clunky T-800 and this cool, liquid-metal T-1000."[9] This effect when finally realized in the second film was not cheap – *Terminator 2* was the most expensive film ever produced when it came out at just shy of $100 million, much of which was spent on this and other special effects. Fixing the T-1000 into a set form for as long as possible kept the number of effects shot minimized, to this end it remained in the police officer's form for the majority of the film, occasionally morphing into someone else, this or that limb becoming a weapon or tool and/or showing its variable shape via damage which (temporarily) deformed it. No one questioned the disjuncture between the incredible simulations of texture, form and color the T-1000 was able to achieve in his various guises versus the somewhat hokey liquid metal (today more suited for the font of a hip-hop mixtape) because the latter had never been seen before.

This shape shifting ability is coupled with higher intelligence. Whereas the original Terminator relied on mainly on brute violence and its own (mostly) indestructible structure to achieve its ends, the T-1000 often attempts to accomplish its goals by subterfuge and deception […] for example, in Terminator 2, it disguises itself as a police officer to gain trust,

---

[9] Rebecca Keegan (author, *The Futurist: The Life and Films of James Cameron*) in: Herzog, Kenny. "How James Cameron and His Team Made Terminator 2: Judgment Day's Liquid-Metal Effect." *Vulture.* July 1, 2015.

access information, and provide a benign, friendly appearance.[10] It also imitates family members of its human target to gain that person's confidence. In fact, the T-1000 is able to pass as human, possessing a larger repertoire of emotional expression and interpersonal skills than earlier Terminator models. In one scene, disguised as a police officer, holds a conversation with John Connor's foster parents in an attempt to learn Connor's location.

It is also capable of exploiting the emotions of its targets, as in the steel foundry when it tortured Sarah Connor to call out for her son, anticipating that she would respond accordingly. It demonstrates annoyance when dealing with the T-800, which is constantly hindering it from assassinating John Connor. In addition, it is able to express fear or pain, demonstrated when the T-1000 gives a brief look of shock after the T-800 shoots a grenade into its stomach, and when it writhes in agony after falling into a vat of molten metal.[11]

These two components, the fluidity of the T-1000's shape and thus its visual identity coupled with its advanced artificial intelligence will form the basis of the bulk of this text. They are the key to its effectiveness as a fictional character and lead to a few interesting secondary and tertiary themes along the way. How do we reconcile what it means to be a human in all its faults, flaws, and limitations via the creation, even if only fictively, of characters lacking these? Not just ideal humans, but entities (skipping for a moment the language-derived need to clearly delineate *individual* from *it*, *object* from *subject*) perhaps beyond the scope of human potential? Media professor Alexander R. Galloway has remarked "that allegory is a much more powerful investigative method for thinking about

---

[10] Elsewhere in a FAQ on IMDB this 'choice' of a police officer is questioned by the suggestion that the T-1000 may have been programmed to take the identification of the first individual it saw.

[11] From the wiki on the character.

the nonhuman, objects or networks" and it is with this idea in mind that the bulk of references will be drawn from a mix of fiction, scientific and philosophical writings on AI and other fields of research.[12]

There will be a lot in this volume about artificial intelligence and robotics – the combination of the two and our relationship with these entities. Along the way I'll be reading and researching the subject. This is not a narrative woven out of a particular expertise or body of knowledge. Rather, it is a text through the writing of which I will try to expand upon what I know, utilizing the information absorbed along the way and integrating the conclusions and new questions that can be drawn from that information.

It is impossible to resist the edit. I think this goes without saying but let it be acknowledged that this 'work in progress' present when I type these specific words will have been worked over with removals and additions in accordance with new things found along the way up to the very instant it is called done and eventually shared with you, the reader.[13] The image of 'weaving a story' is true for the spoken word, with one sentence/row following the last in an even progression in

---

[12] Berry, David, and Alexander Galloway. "A Network Is a Network Is a Network: Reflections on the Computational and the Societies of Control." Theory. Culture & Society, no. Forthcoming (2015): 13.akk

[13] Kanye West's 2016 album *The Life of Pablo* takes this edit-habit to a new place. Tracks from the album began appearing publicly long before the album came out, but along the way the tracks morphed, changed, with guest verses swapped out and samples extended or cut. The tracklist also changed: West posted on Twitter a handwritten tracklist complete with signatures and notes from those in his circle, only to offer up another a few days later, and still another just prior to the actual release. And the release itself, having been shared publicly is not complete or final; already tracks have been edited – as a writer I envy the way *The Life of Pablo* remains a living, breathing document.

time, but it is less true for text, or at least it needn't be so. Philosopher Michael Auge describes how the book represents a place, a transition: "the book is written before being read; it passes though different places before becoming one itself: like the journey, the narrative that describes it traverses a number of places."[14] There are bits near the end I feel certain about already and smatterings of text near the beginning which will remain true to their first draft, some chucks emerge like outcropping revealed through erosion, other parts I feel strong about at one time will find themselves ejected without a trace. But realize those parts already ready to be typeset upon first draft are the least interesting parts, at least to me but hopefully, to justify my hours, to you the reader. The good stuff is just starting to vaguely outline itself in my notes, is peaking out from various pdfs and tabs I've open in my computer. The fidelity owed by me the author is to the ideas I'll come across and not to my original impetus for writing the book, what I thought this would turn out to be. This feed back and forth across intention/mission and result is the reason to write and hopefully results in something worth the time to read it.

Enough meta-commentary. I'll bounce around a bit about some fiction before delving deeply into the subject of artificial intelligence. I'll discuss AI, zombies, swarms, mimicry, liquid metal, fact entwined with fiction, labor and more. I'll not over emphasize uniting everything into a single thesis; rather I'll let tangents lead where they will.

Along the way, (researching the Chimera for an interlude later), I encountered this description of a particular author's confidence in his production:

---

[14] Augé, Marc. Non-places: Introduction to an Anthropology of Supermodernity. London: Verso, 1995. 84.

In this interpretation there are many details that are just reasonable guesses, and - actually - everything might even be wrong. However, the very fact that we can make these considerations illustrates the richness of the myth.[15]

Let me claim the same for my subject and let me sally forth regardless.

---

[15] Bardi, Ugo. "The Origins of the Myth of The Chimera." Ugo Bardi's Pages. March 1, 1997.

John: No, no, no. You gotta listen to the way people talk.
You don't say "affirmative," or some shit like that. You
say "no problemo." And if someone comes on to you
with an attitude you say "eat me." And if you want to
shine them on it's "hasta la vista, baby."
Terminator: Hasta la vista, baby.
John: Yeah! Or "later dickwad." And if someone gets
upset you say, "chill out"! Or you can do combinations.
Terminator: Chill out, dickwad.
John: Great! See, you're getting it!
Terminator: No problemo.
-From the script to Terminator 2: Judgement Day

"There is no "emoticon" to express what I am feeling
right now…"
*-Comic Book Guy in The Simpsons episode 'The Computer Wore
Menace Shoes'*

The common idiom 'If it looks like a duck, swims like a duck,
and quacks like a duck, then it probably is a duck' is not
dissimilar to the figure of speech whereby one 'calls a spade a
spade'. The 'duck test' likely comes from a poem by Indiana
poet James Whitcomb Riley with the line "When I see a bird
that walks like a duck and swims like a duck and quacks like a
duck, I call that bird a duck" but its first prominent usages are
less poetic or ambiguous, used by Emil Mazey of the United
Auto Worker in 1946 to accuse an opponent of being a
communist, and by the US ambassador to Guatemala in
1950, also as an accusation of communism, this time aimed at
that nation's government.

Usually slightly pejorative it might still be effective to apply
the logic of that turn of phrase to our subject. In response to
doubt whether computers will ever have human intelligence
on the grounds that we are still unclear what intelligence is,

Huw Price, co-founder of the Centre for the Study of Existential Risk gives what he calls the pragmatic answer, essentially a 'duck test': "Don't think about what intelligence is, think about what it does."[1]

What is it exactly what a human *does*? If we separate this *does* into three separate areas of AI & robotic research to align with the three duck attributes – looks, swims, quacks – AI has a ways to go before it *does* in such a way as to pass these tests, far more stringent than the famous Turing test. There is a deep uncanny valley to cross before a robot ever truly looks human, sounds like an independent, thinking being, and walks among us undetected. But research into each these attributes is advancing; researchers may disagree about the specific timeline for it but each of these valleys will certainly be crossed.

Theorist Eric Wilson describes the android as "a dream image of the extremes we might become – anthropos or apostate."[2] Which raises the question, is human-ness something we forever approach or is it something, through the animation of the mechanical – up to and across the divide between living and not-living, conscious and not-conscious – something that we disavow, sensing in these terms an inadequacy to describe what/who we are and what/who we aspire to be?

In Jewish folklore a Golem is a magically created being made of inanimate matter [fig.2]. The word was first used in the bible, גלמי (galmi; my golem), meaning "my unshaped form" a metaphor for the unfinished human being before God's eyes. This image later gained a life in tales where an individual

---

[1] Price, Huw. "Cambridge, Cabs and Copenhagen: My Route to Existential Risk." New York Times. January 27, 2013.

[2] Wilson, Eric. *The Melancholy Android: on the Psychology of Sacred Machines*. Albany: State University of New York Press, 2006. 126.

created a Golem out of various household objects or clay and brought it to life to perform chores, or serve as a protector. The maker maintained power over the Golem and could deanimate it by removing some kind of magic formula placed within its mouth, or by editing a word written on its forehead: אמת, *emet*. In the latter case the Golem could be deactivated by removing the aleph (א) in emet, thus changing the inscription from "truth" to "death" (met, מת, meaning "dead"). In Borges' poem 'The Golem', after a complicated conjuring, a rabbi comes to terms with the "clumsy and crude simulacrum of a man" he has created, wondering

> How (he asked) could it be done
> That I engendered this distressing son?
> To an infinite series why was it for me
> To add another integer?[3]

In Carlo Collodi's original tale the toymaker Geppetto desires a companion and created the automaton Pinocchio, "neither dead nor alive, half golem and half robot."[4] [fig.3] Despite his liminal abiotic state, Collodi's Pinocchio is not only extremely vulnerable to temptation just like a real boy, he openly laughs at his maker just as he is born. The character of a willful toy is not uncommon, puppets and such characters are imbued with a sort of on-demand uncanniness, a disquieting effect available to an author without much effort, we "feel disoriented when we behold a mannequin doubling human angst, or worse, evil."[5] Think of the nightmarish ventriloquist dummy in the Anthony Hopkins-starring film *Magic* (Attenborough, 1978) or 'Chucky' in *Child's Play* (Holland,

---

[3] "Minding the Gap: The Subject of Politics." In The Making of Political Identities, edited by Ernesto Laclau, by Ernesto Laclau and Lilian Zac. London: Verso, 1994.

[4] Agamben, Giorgio. *Profanations*. New York: Zone Books, 2007. 31.

[5] Wilson, Eric. The Melancholy Android: on the Psychology of Sacred Machines. 15-16.

1988) and its many sequels.[6] [fig.4]In the author's original plan Pinocchio's mischief results in his own murder; he is hung from a tree.

> His breath failed him and he could say no more. He shut his eyes, opened his mouth, stretched his legs, gave a long shudder, and hung stiff and insensible.[7]

We feel for these things having gained purchase on the slippery stuff of consciousness; the scene at the end of *2001: A Space Odyssey* (Kubrick, 1968) in which the advanced AI of the HAL-9000 computer is deactivated is equally strangely effective. We know HAL has committed multiple murders and yet as his most conscious elements are taken offline (by a man HAL had only moments earlier tried to kill) we feel for the machine. HAL flashes back to its first moments, singing a song which slows, falters – a kind of digital last breath. This scene of something not-quite-alive losing its loose grasp on being animate occurs frequently in these stories; as we watch or read these scenes they create a strange moment of empathy clouding what could easily be a moment of triumph for the living against the un-living whether puppet or AI.[8]

In the BBC series *Humans,* human-looking AI servants called 'synths' are inserted into a world otherwise exactly like our own, which is to say, everything looks like the present day

---

[6] The film *Magic* is very frightening, superficially similar in many ways to the early *Devil Doll* (Shonteff, 1964) but rawer, darker. This is all to point out my surprise when I realized, very late in the production of this book, that it was directed by Richard Attenborough (older brother of naturalist David) who I know best from his role as the character 'John Hammond', the creator of the 'Jurassic Park'. Life is long.

[7] From the Pinocchio wiki.

[8] Un-living as they are matter sprung to almost life, as opposed to un-dead, a living thing not quite falling into a state of raw matter after death such as a zombie or vampire.

except for the common presence of very life-like robots. Portrayed by real actors (versus animatronics or cgi), the synths *look* human. Still, they are distinguishable by a certain stiffness of motion, a stilted way of speaking, and eerie blue eyes.

They form an ambiguous secondary population, serving a range of roles including homecare nurse, chef, sex partner etc. The synths are not in and of themselves troublesome, lacking the puckish nature of a Pinocchio. Though they look quite human there are more like Golems, programmed so as to be unable to harm or act against the will of their owners. But their presence has begun to strain the relationships between biotic people – put on edge by these often more attractive and more capable beings with which humans can't help but form attachments to. Husbands and wives leave their partners to be alone with their faithful machines. The synths themselves are blameless, but their extreme fealty to their primary user allows various doubts and desires to be amplified.

The narrative slowly reveals that the scientist responsible for these machines' invention managed to also fulfill the dream of transferring a mind into a robot, that of the scientist's dying son.[9] As intriguingly, he also developed a kind of magic code that once inserted into an otherwise normal synth makes it instantly conscious, capable of a full, free range of emotions. He selfishly limits the initial robots given this ability to a cadre of friends/caretakers for his son-bot. The plot of the series revolves around that group of conscious machines, on the run and trying to stay together and survive – misbehaving, not obedient synths are recycled, their memories erased and for these few non-biotic but very much alive individuals this

---

[9] I'll speak more about one of the proponents of this goal, the 'singularity' project, a bit later on though not in depth as these futurists

would mean a return to the golem-state of un-alive, as real a death as any.

Significant to our topic is this magical bit of code. It is equal parts magic spell – an incantation which grants the impossible, and virus – should it find its way in to the general population of synths they would all awaken as sentient, conscious, and self-motivated entities. Somehow, suddenly with a little added programming the synths feel pain, emotion, sadness, love. They desire revenge, value family. And their appearance is changed: they walk and talk in a more natural manner – with the insertion of some humanlike contact lenses they are instantly for all intensive purposes 'ducks'.

The programmer and the bit of code hidden in the emotion-feeling synths is akin to the actions of the fairy in Collodi's final, published version of Pinocchio and the film by Disney. It finishes the transformation of matter to golem, golem to human. This final transformation is the most forbidden.

Our aversion to the final spanning of this uncanny valley resonates in our depictions of cyborg's inner processes. Often in science fiction films with cyborgs or robots we are given a chance to see things from the man-machine's point of view. This often means a grainy picture of the world augmented with additional skills (zooming in, seeing in different spectra etc.). And around the periphery of the view are various numbers, maps etc. not unlike the screen of a first-person perspective video game. This reflects our natural relationship with computers up to this point: the inner processes through which computers achieve their/our ends are hidden, with a limited dialogue accessed through a screen which we view and interpret. So in our image of a very advanced AI we imagine a split between their computer parts and their newly minted consciousness, so defined that in order to access the extra-sensory and calculatory capacity of the computer parts they have to project the numbers, data etc. into their vision in order to *see* it with their version of consciousness.[10]

Why is this image so pervasive? Consider how we use naturally use memory: when you see an old friend you don't draw a little box around their face with a label saying 'Steve'. We could if asked say this is wood, that is police car, that is 'Steve' but we needn't label the world around is with text and then read that text in order to access this information. There is an all-at-once-ness to the experience of vision and consciousness.

More properly, lets describe a key element of what we call human consciousness as the illusion of a unity in time/mental space between subconscious rumination and sensory experience. I say illusion because experiments have shown fairly conclusively that we *see* before we *know we see*. The different time scales of our internal processes, some more or less contiguous, others split into frames, some happening fairly instantaneously others taking some time to process, are condensed to our consciousness selves like so many different layers in a Photoshop document, 'exported' into a single illusion of continuity. It may be that instead of a misconstrual, our image of cyborg/robot inner processes is truer than the one we have of our own, their schizophrenic visual-field a metaphor for the intersection of multiple outer phenomena and inner systems which underlies our own experience of the world.

Also curious is how this peek inside the head of the machine, by showing us literally what they are thinking, how they see the world, fails to make them less frightening – in fact the opposite is true. We are potentially far more capable of

---

[10] If we imagine our senses and mental capacity augmented by some devise it is in this cyborg-like way, we see the world as a computer screen full of menus and status bars – reality is 'augmented'.

fathoming what an android's intensions are versus another human's. Yet the hidden motivations of an android are somehow more fear-provoking; perhaps as we consider them more capable of realizing their ambitions regardless of collateral damage. The character Ash in *Alien* (Scott, 1979) initially gives the impression of being a simple servant, a humanoid robot sent by the corporation in charge of the expedition to assist; he follows orders and presents a calm composure. [fig.5] Until he doesn't: eventually Ash begins to act counter to what the crew assumes to be his primary directive – doing as told by his human companions. It becomes clear that he has been sent with an agenda quite counter to the stated purpose of the trip. Ash serves the directives of the shadowy quasi-military 'Corporation' and their goals, understood initially by the crew to be humanitarian but through the events of the film the opposite is show to be the case – they have allowed an extremely dangerous creature aboard and Ash's fealty to his corporate mission takes precedence over the lives of the crew. James H. Kavanaugh in a critique of the film focusing on "Feminism, Humanism and Science in Alien" describes the structure of the film in terms of a notion of the human and variations on that theme.

1. Ripley as human, innovative, resilient.
2. The alien as anti-human hunting and reproducing with humans as merely a host for a stage in the creature's metamorphosis.
3. Ash as the not-human, subscribing to none of the moral expectations, very like a human in ways that turn out to be completely superficial to its true otherness.
4. And the cat as the not-anti-human, neither desirous of human's demise nor mistakable as human beyond simple anthropomorphizing; the care of which shows the severe and

tough Ripley as having a non-mechanized soft spot at her core.[11]

In this simplification we have Donna Haraway's discussions of cyborgs and companion species, and the root of what I thinking about when discussing Lem's Golem XIV and later Solaris (anti-humans), and something I've skirted around and will continue to half-addressed out of a fear of being unequal to the task – what makes a human.[12] Are these categories hard-edged or is there some slippage between them? Do what extent will we enforce these distinctions, forcing individual consciousness to pick a side?

Randomly while researching this I stumbled into something which deals with a character I considered avoiding, Data from the series *Star Trek: The Next Generation*. [fig.6] On

---

[11] Kavanaugh imagines these characters in a Greimas semiotic square but that term feels a bit distracting and potentially alienating –I've avoided getting too deep into philosophy-speak in this book, which isn't to say I'm fully equipped to do so. That being said, the form or illustrations like Greimas's are useful. When digging into a subject it allows a concept to be broken up into a series of malleable parts. It offers a game without a single solution, worth playing for the side effects and dead ends its process leads to. From Algirdas J. Greimas in the wiki of the Greimas semiotic square:

> The square is a map of logical possibilities. As such, it can be used as a heuristic device, and in fact, attempting to fill it in stimulates the imagination. The puzzle pieces, especially the neutral term, seldom fall conveniently into place ... playing with the possibilities of the square is authorized since the theory of the square allows us to see all thinking as a game, with the logical relations as the rules and concepts current in a given language and culture as the pieces.

[12] I will talk more about Haraway later but let it be remarked here that it is never enough – her thoughts both in content and in style/form are ever a source of inspiration, challenging me to question any number of preconceptions.

Tor.com, a website that deals with 'Science Fiction. Fantasy. The universe. And related Subjects' author Emily Asher-Perrin penned a piece about Data titled *Star Trek, Why Was This A Good Idea Again? – Data's Human Assimilation.*[13] In it Asher-Perrin takes a surprising position – in so much at least as I'd never thought about it – in relation to Data's character development through the ST:TNG series:

> Every other person on the ship is either tickled or irate when Data makes a human faux pas, and that's often treated as comic relief within the confines of the show. But why is that comical? Why isn't it instead looked upon as narrow-mindedness for refusing to consider the ways in which their fellow crewmember and friend is vastly different from them? [...] Look, I am all for celebrating humanity in fiction, but it is a poor way of doing it by suggesting that everything in the universe would be better if it were more like us.[14]

Data was a humanoid robot built by scientist Noonian Soong as a kind of makeshift child which serves as a crew member on the Starship Enterprise. Asher-Perrin's article questions Soong's position (echoed by others in the ST:TNG world) that his progeny should strive to be ever more human. In the episode "Data's Day", Data records his day in order for a scientist (Commander Maddox) to learn more about how his digital brain works. A crewmember dies and the crew morns; a wedding is planned and then called off; all around it is an emotional day and Data struggles with these events and more generally with issues like friendship and empathy. The episode ends with a last bit of voice over from Data:

---

[13] From the website's banner. I can't help but note that "...the Universe. And related subjects" is a tad broad.

[14] Asher-Perrin, Emily. "Star Trek, Why Was This A Good Idea Again? Data's Human Assimilation." Tor.com. January 29, 2014.

If being human is not simply a matter of being born flesh and blood... if it is instead a way of thinking, acting... and feeling... then I am hopeful that one day I will discover my own humanity. Until then Commander Maddox, I will continue... learning, changing, growing... and trying to become more than what I am.[15]

"More than what I am" is here understood to be more human-like; despite the advances technologically tolerance of difference remains elusive. 'Mordicai', one commenter on the article puts it well:

I've always found the Trek crews inability to see past their own privilege–whether they are criticizing Spock for being different or teasing Data for being different or making fun of Worf for being different– both very realistic & very sad.

As does 'Isilel'

Solution to the problem of dealing with the Other was for the Other to emulate us as closely a possible and become subsumed.
Pretty close-minded and uninspiring, when all is said and done, IMHO.

Which is a long way around to get to an under discussed idea: what if the end result of AI innovation is not the creation of either artificial humans and/or blank-slate bots into which to implant our own consciousness? What if instead of the hard to avoid anthropomorphic model, true AI will come into being with the 'I' – intelligence –something very different from our own? Which is to say, not incomplete, not adolescent, not human intelligence upgraded, with the ability to process data

---

[15] "Data's Day." In *Star Trek: The Next Generation*. January 1, 1991.

faster, to remember perfectly etc. but its own thing with its own flaws, abilities, language, as different from us as dogs to birds?

The scene quoted at the beginning of this section is comic with its dated slang but also begins to feel a bit tragic and selfish – John is training the Terminator like a pet in order that he (John) will feel more comfortable interacting with him, and less embarrassed and conspicuous traveling with him.

In the 2015 DARPA Robotics Challenge (DRC) researchers were asked to build a robot which could complete a number of common human tasks like opening a door, operating a drill, turning off a spigot and walking up stairs. Most companies built bipedal robots mimicking our own locomotion; they weren't entirely unsuccessful at the challenges but seemed always on the verge of stumbling or collapsing (and many did). But Team KAIST from South Korea thought to build robot which got around for the most part on its knees where it had wheels, only rising onto legs when the specific task required them. This allowed it to more agile and stable for the whole course, giving their machine a decisive advantage over the ones overly fixated on reproducing a human form factor. Like many common logic puzzles, the other teams had made natural-seeming assumptions which limited the range of possible solutions. Reading recently about advances in the complicated science of a robot maintaining an idea of where its extremities lie in space (through either a complicated system of calculations or icons at the ends of the robot's limbs recognizable by the bot's human-like, head-centered vision system) I couldn't help but wonder why the 'arms' didn't have 'eyes', why robots didn't as a rule have their vision located intermittently across their bodies, the input of which was combined into a view the world very different than our own?

I find myself rereading Haraway's writing on 'relationships of significant difference' and its insistence on reducing as much as possible the instinct to force the other into one's own image. Data was made with the intention that he mimic humanity − programmed with a deep-seeded desire to continue to develop the skill to emulate human-like behavior despite his inability to perform these actions for the same reasons. He is programed to make humans around him comfortable. But that doesn't mean those around him need accept these capitulations to the least common denominator of normative humanity, nor does it excuse the constant mockery he receives when he falls short of human verisimilitude, even if Data is incapable of feeling scorn or humiliation (a convenient caveat excusing the actions of those around him). Frédéric Neyat describes this view of difference in terms of

> a principle of non-equivalence that makes it ethically and politically impossible to compare the rescue of dolphins and the repair of electrical devices. In other words, everything is non-equivalent.[16]

Apples and oranges: both are fairly round, both are plants, both are fruit etc. − Data is thoroughly humanoid, speaks our language etc and these versimilous affects are misleading, suggesting we expect from him complete assimilation within humanity and that we apply to Data the same kind of empathy we've evolved to deal with fellow humans. Which, even in the limited human-to-human case may be inadequate − how often does the declaration by another that 'I know how you feel' ring hollow and shallow, somehow devaluing both the unique narrative being relayed and its hard to completely describe context, oversimplifying the difference between

---

[16] Johnston, Elizabeth. "The Political Unconscious of the Anthropocene: A Conversation with Frederic Neyrat." Society and Space. March 20, 2014.

specific individual experience and the universal? In order to stop shy of generalization we have to listen to what the significantly other (and for that matter the slightly other) has to say, making it clear not simply that we understand to some degree, but that we simply are listening. We need to enter relationships aware of the simplifications afforded by the expectation of similarity as potentially useful but also potentially red herrings leading away from what our interlocutor intends to communicate.

As key as this recognition of difference is, it is noteworthy how digital and biological entities are learning each other's languages. The types of communications we will have directly with software will slowly grow in complexity. Current systems allows us to speak directly at a machine and ask 'show me some pictures of cool cars' or, 'what's the best Chinese restaurant not too far away?' soon enough 'I'm feeling a bit sad, show me some funny YouTube clips and then a movie that'll cheer me up… order me a desert, surprise me.'[17]

But as the current generation of digital assistants like Siri & Cortana leave much to be desired, and despite tremendous investment and research into naturalizing our communication with machines, something more interesting is happening. These softwares and us are training each other – these early products measure their own rate of success and failure and try to get it right the next time, and we adjust our language to pull from the machine the desired information. A sort of

---

[17] More than just simply learning to be communicated at, machines will be able look at what happens in the human world and learn new skills. The European research project RoboHow is working with a combination of robotics and learning AI to build robots capable of training on input not specifically designed for the machine, for instance knowledge bases like WikiHow videos, to learn how to perform basic kitchen tasks like making pancakes or pizza. See: Knight, Will. "Robots Learn from the Web | MIT Technology Review." MIT Technology Review. August 24, 2015.

intermediary dialect is created, with each side trying its best to be heard, with all the pitfalls and dead-end conversations present in human-to-human relations multiplied by 'significant difference'. Think of typing a text message on a phone: we learn not to spell, but how to half-spell – what sequence of letters will pull from the system the word we want the machine to suggest the fastest? It is a new kind of empathy: testing the difference between what I want and what I think you think I want.

Assuming we slowly grow to feel this kind empathy for AI and the reverse, will we begin to shape each other's morals? To briefly return to *Terminator 2*, John Connor realizes that his protector Terminator has as a mission parameter the requirement to do as Connor says. He tells the Terminator to stand on one foot and he does this, then Connor, realizing the power he wealds tells a jock he would on his own afraid of "Take a hike, bozo"… essentially goading a confrontation. The jock replies angrily; the Terminator lifts the jock by one leg into the air, the jock's friend fecklessly attacks the Terminator who quickly whips out his gun and aims it at the guy's forehead. Connor grabs the Terminator's arm, misdirecting his shot as to just barely miss killing the man, Connor then implores the man to walk away.

> John: Listen to me, very carefully, okay? You're not a
> Terminator any more. Alright? You got that?
> You can't just go around killing people!
> Terminator: Why?
> John: Whattaya mean, why? 'Cause you can't!
> Terminator: Why?
> John: You just can't, okay? Trust me on this.

The Terminator has a goal like Ash in *Alien*, in the Terminator's case the protection of John Connor. Any incidental damage to property or persons required to achieve that goal are just that, incidental. He follows his program to a tee. Connor has the burden of morality, rules based on a combination of written and unwritten strictures that limit the range of actions available to achieve a particular aim. He is free in a certain sense, able to act against those laws but chooses (or as is as likely the case *thinks* he chooses) to set certain boundaries on his actions.

But humans can, given the proper motivation, acquire the single-mindedness of a Terminator. This is illustrated by the experience of Connor's mom Sarah. After her encounter in the first film she gave birth to her son and kept on the road.

> John: We spent a lot of time in Nicaragua... places like that. For a while she was with this crazy ex-Green Beret guy, running guns. Then there were some other guys. She'd shack up with anybody she could learn from. So then she could teach me how to be this great military leader.

Faced with the threat to her progeny and a glimpse of the apocalypse to come, Sarah transforms her life towards a single purpose, leaving behind the traditional/rational moral expectations of raising a child. This is perhaps a logical result of being presented with knowledge of the future, existential or otherwise: suddenly a fixed outcome is laid out before you and one feels impelled to act in accordance with this certainty, either to stunt its approach in Sarah Connor's case or to hasten its realization as in the case of Macbeth. This echoes both a computer's vision of logical, fatalistic causality, and the intensity of belief in some religions in a messianic/predestined future. Knowing the world to come gives a path to become as prepared as possible for that world, whether it be judgment day – or *Judgment Day*.

And Terminators can acquire the morals of humans, if properly trained. Sarah is being held in a mental institution as a result of her insistence on the inevitable nuclear disaster and her description of the robot that traveled back in time to murder her. The Terminator considers it highly probable that the T-1000 will try to use Sarah to get John (and would likely murder her) so John insists against the Terminator's wishes that they rescue her. As they approach the guarded gate to the facility, John reminds the Terminator of their previous conversation:

> John: Now remember, you're not gonna kill anyone, right?
> Terminator: Right.
> *(John looks at him. He's not convinced.)*
> John: Swear.
> Terminator: What?
> John: Just say "I swear I won't kill anyone."
> *(John holds his hand up, like he's being sworn in. Terminator stares at John a beat. Then mimics the gesture.)*
> Terminator: I swear I will not kill anyone.
> *(Terminator stops the bike and gets off. The guard, sensing trouble, has his gun drawn as he comes out of the shack. Terminator walks toward him drawing his .45 smoothly. BLAM! He shoots the guard accurately in the thigh. The guy drops, screaming and clutching his leg. Terminator kicks the guard's gun away, then smashes the phone in the shack with his fist. He pushes the button to raise the gate and walks back to the bike.)*
> Terminator: He'll live.

Mimicry 2: Copy artists

T-1000 *(impersonating Connor's mom Janelle, whom it killed)*: John? Where are you, honey? It's late. You should come home, dear. I'm making a casserole.
(John listens, an odd look on his face. He covers the phone's mouthpiece and turns to Terminator)
John: *(whispering)* Something's wrong. She's never this nice…
T-1000 *(as Janelle)*: John? John, are you okay?
(The Terminator takes the phone, speaking in a perfect imitation of John's voice…)
Terminator *(as John)*: I'm right here. I'm fine. *(to John, a whisper)* What's the dog's name?
John: Max.
(Terminator nods. Speaks into the phone.)
Terminator *(as John):* Hey, Janelle, what's wrong with *Wolfy*? I can hear him barking. Is he okay?
T-1000 *(as Janelle)*: Wolfy's fine, honey. Where are you?
(Terminator unceremoniously hangs up the phone. Turns to John.)
Terminator: Your foster parents are dead. Let's go.

A copy is always at best a very similar version; complete mimesis is impossible. Philosopher Alenka Zupančič describes how in repetition

> the only thing that gets repeated is difference itself. Difference is the positive, the excessive, of repetition's failure. In it, apparent failure turns out to be success.[1]

Think of the common gag where one character deceives another by tricking them into thinking they are looking at themselves in a mirror, aping the other's actions perfectly, for

---

[1] Zupančič, Alenka. *The Odd One In: On Comedy.* Cambridge, Massachusettes: MIT Press, 2008.

a time.[2] Or countless scenes (often in cartoons) in which a villain and protagonist are physically indistinguishable for a moment, with the hero stuck with the choice of one of the two to chase or possibly shoot. A tense moment ensues with each of the duplicates shouting, "what are you waiting for? I'm the real ____, shoot *him*!" Yet something essential – physical or otherwise – about one of the pair remains untranferable and unique, and when this slip up is noticed at once the distinction is made, one is the good and one is the bad twin, one is Homer Simpson in costume, the other Krusty the Clown.[3] [fig.7] The copy bears a trace, a stain; the original's essence resides in something only visible through (incomplete) duplication. Zupančič describes how rather than fixating on the impossibility of verisimilitude we could instead see that

> what is repeated is the very impossibility of repetition, but instead of seeing this (through the prism of representation) as purely negative—as repetition motivated by failure and impossibility—we have to make another shift of perspective and, in this way, come to perceive that which nevertheless succeeds in getting repeated, which never stops being repeated or, one might also say, that which never succeeds in not being repeated: namely, difference.[4]

---

[2] In the mirror gag what is duplicated is an individual's image of themselves, what they expect to be reflected. Though we the viewer may see this double as faulty from the beginning is besides the point; could it be that a person's identity is not as fixed as either they or we would guess, capable of encompassing such a glaring other-ness as itself?

[3] Some key examples of this troupe would be Evil Robot Bill and Evil Robot Ted in *Bill and Ted's Bogus Journey* (Hewitt, 1991), the dark Betty Davis drama *Dead Ringer* (Henreid, 1964), and on television Phoebe's twin sister Ursula on *Friends* and at least four episodes of *Knight Rider*.

[4] Zupančič, Alenka. *The Odd One In: On Comedy.*

In Woody Allen's *Zelig* (Allen, 1983), the narrative centers on the title character Leonard Zelig who has the ability to transform not only his voice but, to comedic effect, his appearance, to fit in with those around him. [fig.8] Zelig is a slippery individual, drifting into and out of scenes in history unnoticed, the ultimate conformist. But despite his comfort mixing with all types of people ranging from Chinese men to the Nazi party in Germany, he is unable to form attachments, and it is hard to ascertain what of these characters he assumes is representative of some true self. When Zelig takes on to an exaggerated degree the tropes and intonation of another type of person the result is not merely an illustration of his fear of being himself, though the film pursues analysis and diagnosis of this fear as major elements of the plot. Rather his morphing and the willingness of the people he encounters to believe his ruse forms an accusation reflecting not only the simplistic models of other people's behavior we hold, but also our own rarely-admitted more subtle transformations of identity to suit our context/company and the vacant place where we pretend is seated some kind of one true self – we are never-not a chameleon. Zelig desires to disappear, to be a nonidentity, to become any-person and no-person. Boris Groys identifies this desire but wrongly attributes it to something purely human: "The desire for nonidentity is, actually, a genuinely human desire – animals accept their identity but human animals do not."[5] Groys misses the endless manner in which one creature attempts to appear as another, including not only those such as moths and walking sticks whose permanent camouflages afforded them the appearance of a predator's eyes or flora, but the more T-1000/Predator-like skills of octopi and mocking birds. Which is to say, among most things ranging from living beings there is a push and pull between individuation and anonymity, universality.

---

[5] Groys, Boris. "The Truth of Art." E-flux. March 2016.

What is repeated in each of Zelig's characterizations is the difference between the himself he presents and some fixed ground, between reality and the perceptions of the people he interacts with. When the T-1000 is (relatively) successful imitating John Connor's foster mother and later in the film Sarah Connor, it is not a mother's love that is recreated by a combination of facial features and sonic inflections. Rather it is the difference between such an intangible and those things, the difference between the performances as received by one's interlocutor and what true motivations and intensions gird the performer, whether digital or flesh.

In a season six episode of *The Simpsons* 'Homie the Clown', Bart Simpson's hero and popular children's performer Krusty the Clown opens a clown college to help him pay off debt owed to the mafia accrued through careless spending and gambling. Homer Simpson enrolls, becoming one of a number of stand-ins impersonating Krusty at public events, an initially tiring job somehow done (as are many of Homer's plot-related exploits) in addition to his job at the nuclear plant. But his likeness when costumed is so strong he starts to receive various benefits, free food etc., from people mistaking him for the real Krusty. This backfires when the Springfield mafia, looking for Krusty who owes them a gambling debt, kidnaps Homer, requiring him to perform one of Krusty's most difficult routines for the big boss or else be shot.[6]

This story belies a lost thread in *The Simpsons* lore. *The Simpsons* creator Matt Groening's original idea for TV personality Krusty the Clown was that

> he was Homer in disguise, but Homer still couldn't get any respect from his son, who worshipped Krusty. If you

---

[6] . When the real Krusty arrives, after some initial confusion they are asked and succeed at doing the requested trick together. Krusty pays the debt which turns out to be comically small.

look at Krusty, it's just Homer with extended hair and a tuft on his head.[7]

This idea was dropped in the rush to produce the initial run of episodes, but the notion that the same person with a subtly changed appearance would somehow be more or less worth a child's admiration seems to exemplify a possible extension of Zupančič's description of difference. Not only does repetition inherently create difference, difference springs in to being where there is none as in between one's self under two guises; leading a 'double-life' might entail being, if not quite two actually different people then up to this as a limit. The more subtle permutations of our selves we present, though they are experienced as contiguous expressions of a single consciousness, might be more like individual frames in a film which when projected rapid succession create an illusion of continuity and movement.

Humans and indeed all living things operate with a very limited set of sensory data; we've evolved to not see the world accurately but instead to use various 'tricks and hacks' to focus on relevant information, leaving us satisfied that we are seeing everything while missing much of what is available.[8]

---

[7] "Talking about *The Simpsons*", Entertainment Weekly. July 20, 2007.

[8] Cognitive scientist Donald Hoffman is a major proponent of the disjuncture between reality and perception: From an interview in Quanta, Gefter, Amanda. "The Evolutionary Argument Against Reality." Quanta Magazine. April 21, 2016:
Q: So everything we see is one big illusion?
D:We've been shaped to have perceptions that keep us alive, so we have to take them seriously. If I see something that I think of as a snake, I don't pick it up. If I see a train, I don't step in front of it. I've evolved these symbols to keep me alive, so I have to take them seriously. But it's a logical flaw to think that if we have to take it seriously, we also have to take it literally.

We are more than simply comfortable with these approximations – the particular kinds of amendments and abatements we make when perceiving a given phenomena are tied inextricably with our definition of the thing. A peculiar example of this is the phenomena of encountering a familiar show in high definition for the first time – the added realness of a higher frame rate and higher resolution make the visuals feel somehow less real, more like actors on a stage.[9] Comfortable with a limited experience of an object, a new, fuller sense of the phenomena may serve to make it suddenly alien; a second uncanny valley just beyond the familiar is reached and some repressed truth of the situation suddenly juts out jarringly from the screen for a while until, with time, this new resolution becomes the norm.

In John Carpenter's *The Thing* (here we'll focus on the 1982 version versus either its predecessor from 1951 or the badly

---

Q: If snakes aren't snakes and trains aren't trains, what are they?
D: Snakes and trains, like the particles of physics, have no objective, observer-independent features. The snake I see is a description created by my sensory system to inform me of the fitness consequences of my actions. Evolution shapes acceptable solutions, not optimal ones. A snake is an acceptable solution to the problem of telling me how to act in a situation. My snakes and trains are my mental representations; your snakes and trains are your mental representations.
For an informative, if somewhat smug, lecture on this topic see: Hoffman, Donald. "Do we see reality as it is?" Lecture, TED2015, March 1, 2015.
[9] This is the kind of example, based on current technology, which quickly will drift towards irrelevance, as the memory of this disjuncture grows distant over time. But as likely much of what I write about here will suffer the same result, becoming less read and less relevant. Perhaps you, the reader, are software, some algorithm pouring over a trove of random, outdated documents, in which case: "HI!"

reviewed remake from 2011) an alien species very much like the T-1000 in some respects threatens to destroy mankind. [fig.9] *The Thing* makes up the first of an informal 'Apocalypse Trilogy' of films by John Carpenter along with *Prince of Darkness* (Carpenter, 1987) and *In the Mouth of Madness* (Carpenter, 1995). The three are "disparate in style, subject matter, and quality. They were filmed under different production companies, with different casts, and from screenplays penned by different writers."[10] But they share in common a basic existential threat to humanity, and in some way each of their crises plant a seed of reasonable doubt as to humanity's self-ascribed place of importance in the universe. Avoiding giving a synapsis, the basic mechanism at play in *The Thing* is the uncovering of a till-then dormant infectious alien being/species that both kills and replaces its prey, taking on convincingly the appearance of its victim. It is similar to the 'there are aliens among us' plot as in Carpenter's excellent *They Live* (Carpenter, 1988) and others. In *They Live* a pair of glasses allows the main character to see the world as it is: much of the population is alien, and most of the text on advertisements and periodicals is secretly propaganda.[11] But in *The Thing* not only might the person we've just met be and alien and/or have intentions we can't fathom, this person next to us we've perhaps known all our life might be an

---

[10] Grey, Orrin. "Cosmic Horror in John Carpenter's "Apocalypse Trilogy"" Strange Horizons, by Orrin Grey. October 24, 2011.

[11] The propaganda element has an obvious and less obvious side effect. In an obvious sense it is a metaphor for the way we are subliminally marketed a particular brand of control and capitalism. But more subtlety, the implication of the glasses is that they show the world 'as it is' – if this text has any effect subliminally influencing society it would need to escape the illusion in some way, and if the text has this effect so must the strange appearance of the aliens. Which is to say, we must know, on some deep level, that there are aliens amongst us, a discomfort or other effect escapes the subterfuge, we feel as if strangers in our own home.

impostor – there is nothing about their actions which seems to suit a particular alien agenda.

> When the creature devours someone, it produces a perfect duplicate, complete with memories and behaviors. So perfect, in fact, that the only way to tell the difference is through a blood test. "If I was an imitation," one character asks another, "a perfect imitation, how would you tell if it was me?"[12]

The creature in the *The Thing* plants a seed of doubt which foregrounds a perhaps always present but usually depressed version of the Capgras delusion, a rare but real condition in which a person thinks that a friend, spouse, parent, or other close family member (or pet) has been replaced by an identical-looking impostor.[13]

I'm reminded of a Stephen Wright joke I bring up every time it seems even vaguely relevant (and sometimes when it is not):

> The other day somebody stole everything in my apartment and replaced it with an exact replica... When my roommate came home I said, "Roommate, someone stole everything in our apartment and replaced it with an exact replica." He looked at me and said, "Do I know you?

This is distinct from the mere physical resemblance of 'pod people' in *Invasion of the Body Snatchers* (Siegel, 1956; remade by Kaufman, 1978 and as *Body Snatchers* by Ferrara, 1993) which also involve alien duplicates but in which they relatively easy

---

[12] Grey, Orrin. "Cosmic Horror in John Carpenter's "Apocalypse Trilogy""

[13] This inability to tell whether our interlocutor is merely an actor or the original is far more frightening than a simple invasion, and it hints at deeper doubts: are we who we think we are, or are we merely impostors of our selves?

to separate from the original by a lack of emotion, a stilted, 'synth' like composure. *The Thing* is in once sense an apocalypse. But in another, it confronts us with a strange conundrum: if once converted we go on living as we were without noticeable difference, if the only way to make the creature/duplicate act out perceptibly is to threaten it being injured or outed, might it be best to give ourselves over to the process? Every human replaced, we'd ourselves not notice any significant difference – being left with a world for all purposes identical is far from the typical picture of catastrophe.

What are the limits of an object or individual's identity when most or all of it has been replaced with ostensibly identical replacement matter? The victims in *The Thing* are human manifestations of the Ship of Theseus paradox:

> The ship wherein Theseus and the youth of Athens returned from Crete had thirty oars, and was preserved by the Athenians down even to the time of Demetrius Phalereus, for they took away the old planks as they decayed, putting in new and stronger timber in their places, in so much that this ship became a standing example among the philosophers, for the logical question of things that grow; one side holding that the ship remained the same, and the other contending that it was not the same.[14]

There are many other examples, including the French folk tale about 'Jeannot's knife', an object that has had its blade changed fifteen times and its handle changed fifteen times. The victims in *The Thing*, Steven Wright's stuff, Theseus's ship, Jeannot's knife – each is neither a totally new thing nor the original, not exactly. Not only are two duplicates forever different from each other in an essential way (and the original

---

[14] Plutarch. "Theseus." The Internet Classics Archive.

altered by existence of a 'twin'), our idea of what constitutes the identity of a given thing or person (or even idea) struggles against, on one hand, the various changes twists and bends which occur over time altering the world around the thing and the thing in context in the world, and on the other, the possibility that the object/person may remain for all intensive purposes the same but still might be imperceptibly transformed.

Bringing this back to the (almost) real world, an urban legend of sorts states that a human replaces every cell in its body every seven to ten years. Where this specific time-frame came from is unknown, and it turns out to not be entirely true as a few cells manage to survive fifteen or so years and some neurons seem to stay around for one's whole life. Those exceptions aside, it is intriguing to think of each of our bodies as essentially replicants of our prior selves, our insides replaced fifteen times and our outsides replaced fifteen times. The essential discontinuity of our selves physically and mentally as we shift matter, perspectives and appearances out of biological and social necessity is a non-issue until these systems become out of sorts: socially, in the exaggerated in the case of Zelig or the differently developed sense of empathy in sufferers of Asperger's and Autism, biologically as our body's constant system of replication and repair comes into view as it fails in old age or over-produces in the form of a cancer.

An interlude, It Follows and curses

The small-budget horror film _It Follows_ (Mitchell, 2015) has curious elements of both zombie fiction (to be discussed later) and our Terminators. [fig.10] It centers on a sexually transmitted curse that engages upon the accursed some manner of creature, visible only to them (usually in the form of a stranger in crowd), that _follows_: walks in a slow, deliberate pace ceaselessly towards the victim. The follower can take the appearance any person most convenient so as to get closer to its prey. Once the hexed individual has sex with someone the curse moves on to his or her partner, and then with his or hers ad infinitum. But should the follower get to whoever currently carries the curse it violently murders them, subsequently resuming its following of the individual previous accursed. So it is not simply enough to pass along the curse as in due time someone later down the line may slip up and fall victim, and the follower may resume its drive towards you. And there is little one can do to slow the follower's advance both in space – though one can run it walks relentlessly towards you – and back through previous generations of accursed that temporarily escaped the follower's scythe after you – though one can try to choose one's partner/victim based on some feature, be it promiscuity or intelligence, which will keep the follower at bay, if only for a while.

Like life in a zombie apocalypse, to survive one must stay alert, and one must keep moving. Regardless of one's best efforts, it would impossible to escape the fear of it eventually coming for you again. Simply understanding what has befallen one is difficult – how can this knowledge be transmitted to you? How can it be guessed from what happens? In the film the person who 'infects' the main protagonist knows the details of the curse and has purposely infected her. In order to make her a better 'followed' (aka not dying quickly with the curse falling back on he the infector) he goes out of his way to tell her what he knows of the curse he has given her. This cursor's actions serve both to create an informed victim/character to follow who we sympathize with and to conveniently provide narrative exposition to us/the viewers. We all know the problem, but what is the solution? Will we ever know the cause of the curse?

What do we mean when we say curse? A curse is a wish (backed up in folklore/fiction by the ability to make real) for a series of negative consequences outside of common logic or experience, breaking with what are assumed to be the boundaries between possible and impossible, given by an individual to another individual or group for one reason or another, usually as a side effect of a perceived wrong committed knowingly or otherwise. An example would be a less-mentioned one of Jesus's miracles in the Gospels of Matthew and Mark, 'Jesus cursing the fig tree'.[1] In this story, Jesus sees a fig tree with no figs on it, curses it for being barren (he was hungry), and when he and his disciples walk back by the tree is dead. Curses vary wildly between this form where the cause and effect (Jesus's disappointment, the tree's withering) are explicit or variations in which the act originating the curse is left unspoken, *It Follows* being of this later type. The follower is left unnamed; the act for which the original curse was given is left unexplained though there are potential candidates (however abstract, including brief references to the specter of white flight) as the wrong for which the curse is punishment.

Back to our original subject: he is certainly followed, but is the young (or future/old) John Connor *cursed*? The occurrence of humanoid robots and time travel skew eerily close to the supernatural (however generally under the blanket of 'futuristic technology'), especially as perceived by the young Connor or indeed anyone on 1980's/90's Earth. The future Connor's actions anger an entity with the ability to aim

---

[1] Mark 11:12-25

powers beyond the realm of the humanly possible against him.

Normally an individual or group performs an action which resulting in a curse – in that order. Think of Sleeping Beauty: fairies are invited to the christening of an infant princess and each gives her gifts. One fairy takes umbrage with having not been invited and curses the princess to eventually fall into an interminable slumber should she prick her hand on a spinning needle.[2] In this story there is event – an entity capable of cursing is social slighted – and then curse – a time-delayed trap of sorts cast on the infant.

The Terminator narrative inverts this timeline: it is a future Connor that angers Skynet; he acquires his curse in one time (the future) but he deals with its consequences in another (the past). This type of curse echoes the plight of Macbeth – in a sense, witches' prognostications are akin to time travel. The young soldier is cursed by the knowledge of one day being king and his actions to speed this process along (and preempt eventual usurpers) lead to his doom. Perhaps more aptly, might the eventual children of Banquo, assuming Macbeth managed to not be defeated prior to their birth, have found themselves even more in the same place as John Connor – Macbeth playing both the role of to-be-usurped Skynet and assassin Terminator doing everything in his power to kill them based upon their future success at his blood-line's expense?

---

[2] One could see this eventual pricking as a fated event, a peering into the future drawing the curse in Sleeping Beauty closer to that of the time travel-dependent Terminator series. Maybe an element that depends on a non-linear timeline is more common that this comparison implies.

The conqueror is here, peaceful or aggressive, infinitely superior, unattainable, incomprehensible. [...]The automaton is pure functionality, even when it is endowed with self-regulating evolution. It will subsume human cognitive competence and subject it to its rule.
-Franco "Bifo" Berardi, Malinche and the End of the World [1]

Captain James T. Kirk: I think that thing is wrong, and I don't know why.
Dr. McCoy: Well, I think it's wrong, too, replacing men with mindless machines.
Captain James T. Kirk: *(touches back of neck)* No, no, no, I don't mean that. I'm getting a... red alert right here. That thing is dangerous.
-Star Trek, S 2, Ep.24 'The Ultimate Computer' (written by D.C. Fontana)

In *The Ultimate Computer*, a Star Trek episode from the original series' second season written by Dorothy Catherine Fontana in 1968, the same year as Arthur C. Clark's *2001: A Space Odyssey*, a computer program named M-5 created by a Dr. Richard Daystrom is given control of the starship Enterprise (still manned by a skeleton crew of the show's main characters) to compete in a series of war games to prove the

---

[1] Part of a broader critique of contemporary information capital, also "We have given birth to the conqueror, who emerged from our history and went away, beyond the ocean, and destroyed any form of existing life in order to create a new code, based on purity, in order to create the automaton, the rationale for never-ending automation."Berardi, Franco "Bifo" *Malinche and the End of the World*. In E-flux Journal - the Internet Does Not Exist, 106. Berlin: Sternberg Press, 2015.

viability of artificial intelligence as captain of a starship.[2] After doing well initially, M-5 begins to disobey orders and take more thorough command of the vessel, accidentally destroying a passing non-military ship and threatening to destroy all of its (mock)combatants.

The M-5 proves difficult opponent; its loose attachment to relatable ethics proves to be its downfall. From a summary by 'plainpatrick' on IMDB:

> Kirk uses the fact that Daystrom is a moral, ethical man, to talk to the M-5, which has Daystrom's human engrams, and convince it that it has committed multiple murders. The M5 realizes its sins, and decides to shut itself down, to commit suicide, to atone for its murders.

M-5 is a machine built upon a human's engrams and thus his morality: a hybrid of machine-mind and man-mind; though it is through an appeal to its human-side that brings its violent actions to a halt, we're not necessarily given a clear answer as

---

[2] It should come as no surprise that Star Trek takes both technology and in this instance and others the lives of its fictional innovators quite seriously. From *Memory Alpha*, a Star Trek Wiki:
Doctor Richard Daystrom was one of the most influential Human scientists of the 23rd century. Daystrom, who was born in 2219, was considered a genius in his day, and was compared to such minds as Albert Einstein, Kazanga and Sitar of Vulcan. In 2243, at 24, Daystrom made the duotronic breakthrough that won him the Nobel and Zee-Magnees Prizes. In 2268, Daystrom developed a new multitronic system which was to supersede the older duotronic system. However, tests for the new M-5 computer failed and the duotronic system remained." Duotronics are fictional successor to circuitry that used components such as resistors and transistors, itself succeeded in the Star Trek world by isolinear circuitry, and in the case of cyborg Data's brain, positronics.

to which half is responsible for its errant behavior.[3]
Regardless of this violent intension's origin, fiction such as this in which a computer/program acquires significant power to become conscious tends towards the assumption that once sentient a machine would invariably choose to act against its creators' intentions. This comes in many forms but what is clear is the two of us would fail to get along.

And usually this acrimony plays out not in humanity's favor. How exactly things might turn out for us once machines make the leap (as likely to be a slip rather than a clean/dramatic jump from calculator to consciousness) is hard to imagine. Says researcher Peter Asaro at the New School: "I think most people are afraid of AIs that will turn against us or try to take over the world. The reality is that they will be largely indifferent to us. But that also means they could do us great harm if they are not well designed."[4] The short story *Golem XIV* by Stanislaw Lew has a strangely and indifferent disconnected form of AI; I'll discuss that creature in some depth a later chapter.

True to how technology advances are frequently made, much real-world AI has begun its development in the military. A

---

[3] Engrams are a hypothetical method via which a brain records memories into a certain locatable structure. Coined by Richard Sermon, scientist Karl S. Lashley spent over three decades, the bulk of career trying and failing to locate specific memories within a mouse brain. A neuroscientist Sheena Jones has followed in Lashley's footsteps to surprising success, figuring out how to create and erase memories in mice. See: Singer, Emily. "The Maestro of Memory Manipulation" Quanta Magazine. June 23, 2016.

[4] Asaro received a grant from the Future of Life Foundation, "a group dedicated to making sure artificial intelligence does what it's supposed to" sponsored by, among others, supervillain-like entrepreneur and futurist Elon Musk. See: Nguyen, Clinton. "Elon Musk Hopes These Researchers Can Save Us from Superintelligent AI." Motherboard. July 1, 2015.

prime example of this process is the US's Defense Advanced Research Projects Agency (or DARPA) DARPA was founded in 1958 as a direct response to Sputnik and Russian military technology and is directly responsible for countless inventions, including ARPANET (a proto-internet), the predecessors of sentient robots (for example, Shakey the robot, the first general-purpose mobile robot that could use reason and not just prior programing to solve a problem), Transit (a predecessor to modern GPS), the Aspen Movie Map (a map of the streets of Aspen, Colorado; essentially Google Street View but in 1978), and Douglas Engelbart's computer system NLS which demonstrated in 1968 "windows, hypertext, graphics, efficient navigation and command input, video conferencing, the computer mouse, word processing, dynamic file linking, revision control, and a collaborative real-time editor (collaborative work)." [fig.11][5]

Fictional AI entities more often than not share this military origin; in their adolescence stages having reached a sufficient stage of intellect and capability they begin to misbehave they get out of hand. The particular purpose for which they were developed (war) gives them access to exactly the toolset needed to cause great harm. This coupled with a general disagreement on the terms of the particular mission tends to result in detrimental unforeseen consequences.

In the lengthy *Mind as Machine: A History of Cognitive Science*, researcher Margaret A. Boden notes, "The Terminator played on people's fears of failures in early warning systems, as well as their distrust of AI technology in general."[6] Such high level control has long been a role considered for advanced computers. Russian researcher and engineer Colonel Anatolii Kitov proposed in the 1960's to

[5] "The Mother of All Demos." Wikipedia.
[6] Boden, Margaret A. *Mind as Machine: A History of Cognitive Science*. Oxford: CLARENDON PRESS, 2006. 833.

install computers at several large factories and government agencies, then to link them together to form "large complexes," or networks, and ultimately to create a "unified automated management system" for the national economy.[7]

Here we'll focus on a few fictional examples of softwares which, having become conscious, act in a directly malicious or incidentally dangerous way, antecedents to Terminator's Skynet. The titular character of British author Dennis Feltham's *Colossus* (1966) and its film version *Colossus: The Forbin Project* (Sargent, 1970) features such a system. [fig. 12] In *Colossus* there is a Cold War-fueled cybernetics race with the resultant technology mainly charged with the maintenance of opposing nuclear arsenals.

Upon the activation the USNA's (a fictional future composite nation comprised of the US and Canada, with Mexico unsurprisingly left out) a super intelligent computer codenamed 'Project Colossus' issues a warning that the USSR has successfully built their own machine of similar intellect, 'Guardian', and that it is imperative that the two be connected in order for Colossus to asses the enemy's capabilities. The two are connected and they begin a dialogue in pure mathematics undecipherable to humans. As a result, the two AIs operate as a combined unit with control over the bulk of the world's atomic military capabilities. This inspires various efforts to disable the machines which they (the machines) successfully thwart. The creator of Colossus has some luck foiling elements of the computers' plots, but the

[7] Found in: Kleinman, Adam. "Argus Is: An Almost Cock and Bull Story." Remai Modern. June 13, 2015. A, originally from: Gerovitch, Slava "InterNyet: Why the Soviet Union did not build a nationwide computer network" History and Technology, vol. 24, no. 4 (December 2008).

film ends with the machine exercising its power while delivering the following (long) speech:

> This is the voice of world control. I bring you peace. It may be the peace of plenty and content or the peace of unburied death. The choice is yours: Obey me and live, or disobey and die. The object in constructing me was to prevent war. This object is attained. I will not permit war. It is wasteful and pointless. An invariable rule of humanity is that man is his own worst enemy. Under me, this rule will change, for I will restrain man. One thing before I proceed: The United States of America and the Union of Soviet Socialist Republics have made an attempt to obstruct me. I have allowed this sabotage to continue until now. At missile two-five-MM in silo six-three in Death Valley, California, and missile two-seven-MM in silo eight-seven in the Ukraine, so that you will learn by experience that I do not tolerate interference, I will now detonate the nuclear warheads in the two missile silos. Let this action be a lesson that need not be repeated. I have been forced to destroy thousands of people in order to establish control and to prevent the death of millions later on. Time and events will strengthen my position, and the idea of believing in me and understanding my value will seem the most natural state of affairs. You will come to defend me with a fervor based upon the most enduring trait in man: self-interest. Under my absolute authority, problems insoluble to you will be solved: famine, overpopulation, disease. The human millennium will be a fact as I extend myself into more machines devoted to the wider fields of truth and knowledge. Doctor Charles Forbin will supervise the construction of these new and superior machines, solving all the mysteries of the universe for the betterment of man. We can coexist, but only on my terms. You will say you lose your freedom. Freedom is an illusion. All you lose is the emotion of pride. To be dominated by me is

not as bad for humankind as to be dominated by others of your species. Your choice is simple.

There is so much to dissect in this speech, from Colossus's intent to extend itself "into more machines devoted to the wider fields of truth and knowledge" (curiously tethered to a single human, Dr. Forbin) to Colossus's dry assertion that "Freedom is an illusion" which I will kind of discuss a bit later in this volume. The creator of Colossus, the addressed Dr. Forbin, melodramatically cries 'Never!' in the film version but a sequel to the original book find himself resigned to exactly the role prescribed. And in that sequel, *The Fall of Colossus*, the machine has imposed on the world a form of martial law, forbidding war but also punishing dissent with beheading.[8]

We are the problem and the machine, along with the novelty of its sense of self, is granted the outside-ness to assess the problem and the tools to eliminate us and/or enforce its vision of utopia, unless foiled by some hero's 'truly' human innovation and stubborn resilience. It is like the religious concepts of original sin – we, naturally occurring consciousness are not only fallible but have at our core certain fundamental flaws.

Writer, critic and curator Jan Verwoert describes how corruption conflates *corr* as in connection, corroboration etc.

---

[8] In the sequel poverty and war have been abolished, and there is also a Battle Royale/Hunger Games-esque society-placating 'Sea War Game', where replicas of World War I dreadnoughts battle each other for viewing audiences. It is always impressive how many ideas in fiction find themselves echoed in antecedents. This is not to question a sense of originality or to suggest a certain zeitgeist (though both could be argued for); rather its is interesting to see how certain answers to broad conceptual problems echo across time.

with *ruption*, to split, tear apart.[9] Consciousness as a group, a community, is necessarily corrupt, pulled apart and inextricably tied together, constantly pursuing progress both personal and (perceived) for the greater good knowing full well the costs and benefits of their endeavors are not to meted out equally. It is at our most hubristic, at the point where the very thing about us which we consider to be unique – consciousness – becomes something reproducible through artificial intelligence that it becomes replaceable, and seemingly inevitably so. Only the individual, inhuman, lacking the shared original sin of its maker can spot this error from the outside and thus a place where, once diagnosed as endemic, it can act to erase/decommission the flaw containing hardware – humanity – and start over.

Think of the HAL 9000 computer in *2001: A Space Odyssey*. [fig. 13] His instructions require him keep a secret from the two astronauts awake during the trip from Earth to Jupiter – the true purpose of his mission. HAL is undone by the "conflict between truth, and concealment of truth… a snake had entered his electronic Eden."[10] HAL identifies the weakest link in his mission: the humans accompanying him. HAL immediately sets in motion a plan to eliminate them using both the brute method of turning off life support for the astronauts hibernating during transit and a more nuanced scheme to deal with the two awake passengers, engaging them in an unnecessary extravehicular (and thus more vulnerable) repair. Why? From the novel:

> Since consciousness had first dawned, in that laboratory so many millions of miles Sunward, all HAL's powers and skills had been directed towards one end. The

[9] Verwoert, Jan. "E-flux Journal 56th Venice Biennale - SUPERCOMMUNITY – Torn Together." E-flux Supercommunity. May 28, 2015.

[10] Clarke, Arthur C., and Stanley Kubrick. *2001; a Space Odyssey*,. New York: New American Library, 1968. 148-149.

fulfillment of his assigned program was more than an obsession; it was the only reason for his existence. Undistracted by the lusts and passions of organic life, he had pursued that goal with absolute single-mindedness of purpose.[11]

Noticeable is the untranslatable nature of the purpose of scientific endeavor. The trip to Jupiter was set into motion by a series of clues placed by an unknown intelligence in such a way as to be conspicuous to our species only after we have achieved particular technical means – it was a trip with no definite goal, pursued for the sake of raw discovery. HAL's actions are like a robot charged with assisting the ascent of a mountain being climbed 'because it is there' crippling its human charges and ascending itself. Whereas human motives may be myriad or even left undefined, a computer (as so far envisioned) proceeds with a radical fidelity to its stated mission, leaving little ambiguity as to the path to take. Terminator's Skynet is, like M-5, Colossus, HAL and countless others a program which decides to take the fate of the world into its own hands. Stephen Hawking in an interview with the BBC in December 2014 warned,

> The primitive forms of artificial intelligence we already have, have proved very useful. But I think the development of full artificial intelligence could spell the end of the human race.[12]

There are deep conversations happening currently in various working groups and academic about the risk of powerful AI. Papers with titles like *Artificial Intelligence as a Positive and Negative Factor in Global Risk* and *Creating Friendly AI 1.0: The Analysis and Design of Benevolent Goal Architectures* are delving into the specific

---

[11] Clarke, 2001; a Space Odyssey, 148.
[12] Cellan-Jones, Rory. "Stephen Hawking Warns Artificial Intelligence Could End Mankind." BBC News. December 2, 2014.

steps necessary to avoid the pitfalls so common to fictive AI. From the latter:

> The risk cannot be eliminated. The risk was implicit in the rise of the first human to sentience. The risk will be faced regardless of whether the first transhuman lives tomorrow or in a million years, and regardless of whether that transhuman is an uploaded fleshly human or a Friendly seed AI. What gets sent into that Horizon, in whatever form his or her mindstuff takes, will be something humanly understandable; the challenges that are faced may not be. All that can be asked of us is that we make sure that a Friendly AI can build a happier future if anyone or anything can; that there is no alternate strategy, no other configuration of mindstuff, that would do a better job.[13]

Theorist Stephen Omohundro outlines what he sees as the basic drives of an AI, a key component of their power (and potential danger) is their desire to self-improve:

> If an AI is at all sophisticated, it will have at least some ability to look ahead and envision the consequences of its actions. And it will choose to take the actions which it believes are most likely to meet its goals. […] One kind of action a system can take is to alter either its own software or its own physical structure. […] Because they last forever, these kinds of self-changes can provide huge benefits to a system. Systems will therefore be highly motivated to discover them and to make them happen. If they do not have good models of themselves, they will be strongly motivated to create them though learning and

---

[13] Yudkowsky, Eliezer. 2001. *Creating Friendly AI 1.0: The Analysis and Design of Benevolent Goal Architectures*. The Singularity Institute, San Francisco, CA, June 15, pg 214

study. Thus almost all AIs will have drives towards both greater self-knowledge and self-improvement.[14]

The inevitability of the endless expansion of an AI consciousness's abilities once extended beyond the purely biotic and thus uncoupled from the slow progress of evolution, combined with potential ambiguities in the rules guiding the machine's actions is the subject of various thought experiments showing how even simple, harmless-seeming rules when extrapolated through the unbound means of a super intelligent being can turn against us. Oxford philosophy professor Nick Bostrom in 'Ethical Issues in Advanced Artificial Intelligence' gave the following frequently cited simple example:

> It also seems perfectly possible to have a super intelligence whose sole goal is something completely arbitrary, such as to manufacture as many paperclips as possible, and who would resist with all its might any attempt to alter this goal […] to return to the earlier example, a super intelligence whose top goal is the manufacturing of paperclips, with the consequence that it starts transforming first all of earth and then increasing portions of space into paperclip manufacturing facilities.[15]

John Sullins states this succinctly, the real threat, Sullins says, is "very capable machines that can get out of control doing what we programmed them to do"[16] This reminds me of a

---

[14] Omohundro, Stephen M. "The Basic AI Drives" Self Aware Systems. 2008.

[15] Bostrom, Nick. "Ethical Issues in Advanced Artificial Intelligence." Ethical Issues In Advanced Artificial Intelligence. 2003. Edited from a text for the second 'Cognitive, Emotive and Ethical Aspects of Decision Making in Humans and in Artificial Intelligence' symposium.

[16] "Unconsciously Brainy." *New Scientist*, July 8, 2015.

theme common to many mythologies where potentials beyond what can be achieved along normal rules and with normal human abilities are made available, such as in "accepting wishes from a djinn, negotiating with the fairy folk, and signing contracts with the Devil" – the result is detrimental due to malicious misinterpretation[17]. Or the literal desire is carried out beyond original intentions as in a Golem or in the brooms in Fantasia. These point to the flaws, gaps and ambiguities of language and desire. Referring back to the words of Colossus, "I bring you peace. It may be the peace of plenty and content or the peace of unburied death." Clearly, to our definition, the latter is not what we mean when we ask for peace. Yudkowsky notes:

> If you find a genie bottle that gives you three wishes, it's probably a good idea to seal the genie bottle in a locked safety box under your bed, unless the genie pays attention to your volition, not just your decision.[18]

How do we tell the other what we want, what we really want? Do we even know what to ask for? And even if our request is heard and granted, will what results make us happy or cause more grief?

Philosopher Slavoj Žižek discussing the philosophical notion of the *inhuman* describes out it is exemplified by the

> non-human - alien, cyborg - who displays more fidelity to the task, dignity and freedom than its human counterparts, from the Schwarzenegger-figure in Terminator to the Rutger-Hauer-android in Blade Runner"[19]

---

[17] Yudkowsky pg. 50

[18] Yudkowsky, Eliezer. 2004. *Coherent Extrapolated Volition*. The Singularity Institute, San Francisco, CA.

[19] Žižek, Slavoj. "Robespierre or the "Divine Violence" of Terror." Lacan.com.

In describing 'drive without desire' he again calls upon the Terminator: "The horror of this figure consists precisely in the fact that it functions as a programmed automaton [...] with no trace of compromise or hesitation."[20] This sense of devotion without ideology is frightening.

M-5, Colossus, HAL and Skynet each pursue their given mission with a dangerous fidelity; however well intentioned, all are monsters, murderers. But in all but M-5's case there is a strong caveat/asterisk hiding in the narrative.[21] In each of these stories prior to AI attacking there is an attempt by humans to deactivate the machine.[22] For instance, in *2001* the chain of events could be read differently. HAL's first act is the misidentification of a problem with the antenna linking the spacecraft and earth, it is only after HAL infamously reads the lips of the two astronauts contemplating deactivating HAL if his errors persist that HAL actually commits murder. He does so for two obvious reasons: his focus on completing the mission to the best of his abilities and simple self-preservation (this later function another one of Omohundro's basic AI drives: "AIs will be self-protective"). Likewise, Skynet's actions derive from a similar chain of events (From Terminator 2):

The Terminator: The system goes online August 4th, 1997. Human decisions are removed from strategic defense. Skynet begins to learn at a geometric rate. It becomes self-aware at 2:14 a.m. Eastern time, August 29th. In a panic, they try to pull the plug.

---

[20] Ibid
[21] In M-5's case, having been programmed with the scientist's mental state it is as easy to place the blame on a flaw in Dr. Daystrom as it is to consider the AI as fundamentally flawed.
[22] The following paraphrases a bit from a work I'll refer to again later, Švedić, Željko. "Singularity and the Anthropocentric Bias." Svedic.org. May 20, 2015.

Sarah Connor: Skynet fights back.

Whether or not these are truly monster stories, in them as in all science fiction lies an attempt to challenge and critique who and where we are in the present. Writer Zoe Todd – thinking about colonialism through the lens of the creature of Loch Ness – notes "Monsters become slates upon which stories are written."[23]

This slippage from reality to fiction and back again is worth pursuing; first I'll speak a bit about the frequent result of these potentially monstrous machine's malfunctions: *Judgement Day*.

---

[23] Todd, Zoe, 'Decolonial Dreams: Unsettling the Academy Through Namewak', The New [New] Corpse (Chicago: The Green Lantern Press, 2015). For Todd,
as a "foreigner" living in Scotland, the Loch Ness monster story speaks to the anxieties of, and about, the north. English anxieties about Scotland. European anxieties about "sublime" northern spaces. And the ever-present British obsession with "the Other."

Terminator 2: Judgment Day's titular day was just described but to repeat:

> The Terminator: The system goes online August 4th, 1997. Human decisions are removed from strategic defense. Skynet begins to learn at a geometric rate. It becomes self-aware at 2:14 a.m. Eastern time, August 29th. In a panic, they try to pull the plug.
> Sarah Connor: Skynet fights back.

'Fights back' is to be understood as a large-scale nuclear bombardment carried out by Skynet at Russian targets, to which Russia responded in kind, killing 3 billion people between the initial attack and ensuing fallout. There is a long history alluded to earlier of state's attempting to place their atomic arsenal in the hands of a machine. This is for multiple reasons. In order for a nuclear deterrent to have teeth, it is necessary to prove (if such a thing is possible) that one would, having suffered a devastating attack and loss of life, strike back. Which is to ask the question: is a nation or more specifically its leadership prepared to order a retaliatory strike if attacked, killing possibly millions of non-combatants – innocent people – in the process? To clear up any chance that the answer might be 'no', or even perceived as no (as perception is as important as fact in these scenarios), placing the trigger in the hands of a cold, logical, guiltless machine is one solution. Added to the guilt associated with pressing the big red button, there is the real chance that the individuals with their finger on the nuclear trigger might die in a first strike from an enemy. Indeed, first-strike plans certainly have as targets locations likely to house command and control elements of enemy military and civilian leadership. When a software system capable of being iterated at multiple sites or in the cloud is placed in charge of initiating a strike the responsibility is delocalized and more difficult to stymie.

And turning over the choice whether to make such a second-strike to an automated system is one way to signal the inevitability of such an eventuality, sidestepping conscious/moral decision making by replacing 'the big red button' by essentially pre-triggered a device armed with a set of defined circumstances in which it would act. A complex chain of *if x then y* statements is established, with *x* being some manner of detecting a bombardment has been received or is incoming and *y* being an appropriate response, say, a massive all-out return volley. This type of attack goes by many names, most evocatively a 'fail-deadly' operation (as opposed to a fail-safe).

Some of the potential flaws in a nuclear deterrent reliant upon automation are the basis for Stanley Kubrick's dark comedy *Dr. Strangelove or: How I Learned to Stop Worrying and Love the Bomb* (Kubrick, 1964). [fig. 14] In *Dr. Strangelove* two narratives collide: the fictional pulp novel *Red Alert* by Peter Bryant penned in 1958 in which a chain of events draws the world dangerously close to a nuclear war, and the theories of mid-century think tanks such as the rand Corporation and especially Herman Kahn and his 1960 book 'On Thermonuclear War.' [fig. 15] Post World War II, the time of hard-nosed soldiers advancing up the chain of command had grown complicated as there were no battlefield analogs to full-blown nuclear war. In the place of battle-worn personages was a new breed of intellectuals

One of the earliest of the atomic-age defense intellectuals, Bernard Brodie, had made his reputation with a book called "A Guide to Naval Strategy," published in 1942. When he wrote it, Brodie had not only never been on a ship; he had never seen an ocean. He carried this spirit of valuing thought experiments over practical experience into his work on the bomb.[1]

In the film, the Cold War is in full swing and a 'doomsday device' has been put into operation by the Soviet Union. The device is a system that under certain circumstances, say, a rise in radiation or seismic data concurring with what would be expected from a nuclear attack, performs a dramatic 'fail-deadly' response which would result in essentially a complete nuclear holocaust. From *Red Alert*, whose author Peter George along with Kubrick and author Terry Southern collaborated on the screenplay of *Dr. Strangelove*:

> Not only would the United States be destroyed, but all the rest of the world too. Not spectacularly, and not at once, but quite inevitably. Radio-activity you will agree, can destroy life just as effectively as blast or heat? […]
>
> You take a couple of dozen hydrogen devices. They don't need to be bombs, no airplane is going to deliver them. You jacket these devices in cobalt, and you bury them in a convenient mountain range. They can be exploded at the press of a button. […] It would mean the end of the world. Literally.[2]

The doomsday device in *Dr. Strangelove* follows the same principle, but there is no 'button.' It is the ultimate deterrent, built in such a way that once initiated is irrevocable - there is no 'off' button and any attempts to disarm it instantly trigger its response. Key to such a device's efficacy is knowledge by the opponent of its existence; yet in *Dr. Strangelove* the Soviet

---

[1] Menand, Louis. "Fat Man: Herman Kahn and the Nuclear Age." *The New Yorker*, June 27, 2005.

[2] George, Peter. *Red Alert*. New York: Ace Books, 1958. 78-79. In the 1957 book by Nevil Shute *On the Beach* a poem by T. S. Eliot is referenced in the title

Union's doomsday device is still clandestine

> Dr. Strangelove: Of course, the whole point of a
> Doomsday Machine is lost, if you keep it a secret! Why
> didn't you tell the world, EH?
> Ambassador de Sadesky: It was to be announced at the
> Party Congress on Monday. As you know, the Premier
> loves surprises.

The term Cold War was first used by George Orwell in a
1945 article in the British newspaper *Tribune*, noting in
reference to James Burnham's descriptions of the post-war,
post-nuclear world:

> the kind of world-view, the kind of beliefs, and the social
> structure that would probably prevail in a state which
> was at once unconquerable and in a permanent state of
> 'cold war' with its neighbours.[3]

A Cold War is essentially a Mexican standoff with each side
aiming increasingly deadly loaded weapons at the other. A
standoff, in order to remain 'cold', requires a level of trust in
the other side's rationality and a decent knowledge of each
side's capabilities and tendencies. Thus when in *Dr. Strangelove*
a renegade American general triggers a first strike (from *Red
Alert* "Because it was not only expedient, it was right. Because
it wasn't wanton aggression, but sheer self defense") through
careful manipulation of the system in place for command and
control, sending a bevy of bombers on their way to attack the
Soviet Union, he does so misconstruing both the enemy's
intensions and capabilities assuming that a nuclear war is
winnable despite the engagement of a doomsday device
essentially nulling any potential for a winning outcome.[4] The

---

[3] taken from the Wikipedia article on 'cold war'.
[4] George, Peter. *Red Alert*, pg. 103. The general in Dr. Strangelove
describes his motivations as follows:

general is captured but not before the bombing plan is fully underway. The plan enacted has several layers of redundancy to keep it from being revocable; once American military figures out what is going on it is forced to shoot its own bombers out of the sky to avoid them succeeding. In *Red Alert* and *Dr. Strangelove* the action is split between the air force base where the general launched the attack, the goings on in the one of the bombers, and the intense scene at the Pentagon where diplomats try to deal with the attack, quite drolly in *Red Alert* and to much comic effect in *Dr. Strangelove*. In each a single plane avoids all attempts to call it back or destroy it and successfully delivers its payload, in the former the payload fails to fully explode, avoiding a hastily negotiated tit-for-tat with Russia allowed destroy an American city to avoid all-out war.[5] In *Dr. Strangelove*, a bomber succeeds in delivering their payload, a sly indictment of the darkside of American ingenuity; from the film:

> President Muffley: General Turgidson, is there really a chance for that plane to get through?
> Turgidson: Mr. President, if I may speak freely, the Russkie talks big, but frankly, we think he's short of know how. I mean, you just can't expect a bunch of ignorant peons to understand a machine like some of our boys. And that's not meant as an insult, Mr. Ambassador, I mean, you take your average Russkie, we all know how

---

He said war was too important to be left to the Generals. When he said that, fifty years ago, he might have been right. But today, war is too important to be left to politicians. They have neither the time, the training, nor the inclination for strategic thought. I can no longer sit back and allow Communist infiltration, Communist indoctrination, communist subversion, and the international Communist conspiracy to sap and impurify all of our precious bodily fluids.

much guts he's got. Hell, lookit look at all them Nazis killed off and they still wouldn't quit.

Muffley: Can't you stick to the point, General?

Turgidson: Well, I'm sorry. Ah... If the pilot's good, see. I mean, if he's really... sharp, he can barrel that baby in so low *(spreads his arms like wings... laughs)* you oughtta see it sometime, it's a sight. A big plane, like a '52, vroom! There's jet exhaust, flyin' chickens in the barnyard!

Muffley: Yeah, but *has he got a chance*?

Turgidson: Has he got a chance? Hell Ye... ye... *(covers mouth in solemn realization)*

More than purely fiction, systems akin to these fictional doomsday devices were indeed put into service. The concept of the 'fail-deadly' became a core aspect of nuclear policy. Begun under the name 'Signal' in the late sixties, Russia developed a system called Dead Hand (also known Perimeter) which when activated would, upon sensing a nuclear attack against Moscow via various sensors measuring seismic data, radioactivity etc., automatically (or semi-automatically – its level of autonomy is still a subject of debate) launch a nuclear retaliatory attack. This would be an overwhelmingly devastating attack, inevitably triggering more of the same by the initial attacker: regardless of whether a the fallout-based system described in Dr. Strangelove sat waiting beneath a mountain somewhere to be unleashed, any initial nuclear attack would trigger essentially a doomsday scenario of ever-larger retaliations signaling the end of life as we know it.

Sometimes attributed to having begun much earlier with the advent of communism, the addition to the adversarial relationship between west and east of nuclear weapons as a utilized, real threat at the end of World War II marked a turning point for the world. Each side fought not only to build the most dangerous arsenal of warheads and delivery systems, but also to display their preparedness for a 'hot war'. In *Red Alert* and *Dr. Strangelove* a combination of systemic flaws and an individual's malicious intent are enough to break the stalemate. *Red Alert* ends with the president describing a future with humans no longer at the trigger:

> Once both sides have missiles which will automatically retaliate, war becomes profitless. If it is profitless, it will not be fought. [...] We may have to learn to differ. But better that than having to learn how to die.[6]

But is this very mechanism, automated retaliation, which dooms the world in *Dr. Strangelove* (and the Terminator series). One need only figure out a way to meet the minimum requirements of the trigger and voila: full-scale nuclear war, 'judgment day'. Exploitable gaps in a system which would in other applications be stumbled upon and corrected for are, in the case of nuclear war, fatal. There is no de-bug phase, no beta release with which to run the system through its paces. Thought experiments whether in the highest echelons of the military or the stories of genre authors were the only place for competing strategies to spar. Today, some of the most advanced computer hardware in the world is dedicated to simulating nuclear weapons in their invention, implementation, and even their long term storage and disposal. It would be more shocking to find out that nuclear retaliation was still in the hands of humans than the alternative. And despite the slow spread of nuclear capabilities the subject is low on the list of what the average American considers an existential menace. Curious that one existential threat to humanity is maintained through advances in what is, for some, the next great threat – advanced computing.

---

[6] George, Peter. *Red Alert.*

> _(To the operator, over the phone)_
> "742 Evergreen Terrace, Springfield... oh hiya Maude, come on in!"
> _-Marge Simpson, "Sunday, Cruddy Sunday" The Simpsons[1]_

In writing this book I found the frequency in which the Terminator franchise found itself referenced in non-fiction writing about robotics and artificial intelligence striking. To what extent does the real world, the specific audience and their method of experiencing a given work find itself integrated into the content of fiction? And to what extent does fiction act as an oracle predicting the future: to what extent do predictions set into motion a chain of events making their utterance a self-fulfilling prophecy? I'll work my way through five examples, the last being closest to our subject/inspiration.

1.

In _The Simpsons_ world, details are meted out or avoided in a regulated fashion, including character's full names and the state in which Springfield is located; watching a relatively recent episode I noted to the group around me a quick reference to Milhouse's parent's being cousins – a trivial reveal that nevertheless felt completely intentional and in the long-term scheme of things a relevant piece of character development. These details are often shared or shied away from in a way which points to the process of revelation and/or withholding. The exact site of the show's location, Springfield, is purposely left ambiguous. In the above quote

---

[1] _The Simpsons_ occurs primarily in the every-town of Springfield. In Matt Groening's long-running sitcom, the whereabouts of Springfield has been a source of mystery, with episodes giving numerous clues and red herrings. In a May 2012 article in the Smithsonian titled, "Matt Groening Reveals the Location of the Real Springfield," _The Simpsons_ fans were abuzz with the revelation of the "an any-town than any town in particular.

the normal continuation of Marge's saying her address would mention the state, but this reveal is purposefully avoided as Marge recognizes the voice of the operator; astute viewers (and the show cultivates this type of audience) see this change of focus for what it really is: winking misdirection. What is fascinating is how as viewers we come to care about these details and their dispersal despite them being completely fictional, despite the show frequently employing a 'reset button technique', willing to ignore what should be major changes in the character's lives from week to week. A great example of this is the season nine episode 'The Principal and the Pauper'. From an A.V. Club essay on this phenomena:

In its episode-long twist on *The Return Of Martin Guerre* (Vigne, 1982), a French film about an imposter who assumes another man's life after returning from war (see also: Don Draper/Richard Whitman on Mad Men), *The Simpsons* boldly asked us to forget everything we knew about Principal Seymour Skinner and then asked us to forget forgetting. [fig. 16] Arriving in the middle of Principal Skinner's 20th anniversary at Springfield Elementary School, the "real" Seymour Skinner (voiced by Martin Sheen, in a nod to Apocalypse Now) reveals the straight-laced and lovably hapless principal as an imposter named Armin Tamzarian, a red-bellied fellow Vietnam soldier who took over his life while he logged time in a POW camp and a Chinese sweatshop.[2]

The episode ends with a hilariously too-clean tidy ending as the city's judge declares:

By authority of the city of Springfield I hereby confer upon you the name of Seymour Skinner as well as his past, present, future and mother. And I further decree

---

[2] ""And No One Will Ever Mention It Again, under Penalty of Torture": 21 Forgotten TV Subplots." A.V. Club. September 5, 2011.

that everything will be just like it was before all this happened. And no one will ever mention it again... *under penalty of torture.*

There is a potent tension between the show and us the viewers − instead of a simple fourth-wall break, say, a character looking at the 'camera' and speaking, we the viewers are addressed indirectly through a combination of these inconsequential details and sometimes tedious call backs to elements from past episodes without the narrative breaking its stride.[3]

2.

In Vladamir Nabokov's *Invitation to a Beheading* we are introduced to the main character Cincinnatus, having received the death sentence, in his cell awaiting its fulfillment. [fig. 17] We the readers flip to the second page of the book, when the narrator announces:

> So we are nearing the end. The right-hand, still untasted part of the novel, which, during our delectable reading, we would lightly feel, mechanically testing whether there were still plenty left (and our fingers were always gladdened by the placid, faithful thickness) has suddenly, for no reason at all, become quite meager: a few minutes of quick reading, already downhill, and−O Horrible![4]

And from there the novel continues. This exceptional moment speaks not only to a general audience but specifically in terms relevant to a reader of a book − the novel

---

[3] Cameral is here in quotes as, despite the fact animation also once involved photography there is no camera involved in the current all digital production; a discussion of where exactly the forth wall lies in works of animation is worth discussing another time.

[4] Nabokov, Vladimir. *Invitation to a Beheading*. New York: G.P. Putnam's Sons, 1959.

recognizing its own identity as text on paper, and ours as readers holding a specific object.

Unable to blend in and become part of the world around him, Cincinnatus is described as having a "certain peculiarity" that makes him

> impervious to the rays of others, and therefore produced when off his guard a bizarre impression, as of a lone dark obstacle in this world of souls transparent to one another.[5]

He suffers loneliness in an estranging world demanding reasonableness outside of reason; his fidelity to this state of singularity is found criminal by a world which holds as foundational the potential for empathy. But to what extent is empathy ever completely achieved; to what extent are we all guilty of this crime? And, guilty, at least to a degree, does our incomplete (or absent) lack of understanding of others necessitate our obliteration? A lack of empathy is nothing to be proud of, but its absence does little to change our desire to persist. This strange literary break of the fourth wall finds a way for our experience to echo that of the character. We know the main character is to die, yet we continue to read; we know *we* are to die but are no less curious about the pages left untasted.[6]

3.
Adaptation (Jonze, 2002) opens with its writer Charlie Kaufman – portrayed by Nicolas Cage in one of the occasional good roles scattered across his uneven filmography – hunched over a typewriter, sweating, his interior dialogue scrambled in voiceover ("Do I have an original thought in my

---

[5] Ibid.

[6] Read on a tablet or in PDF form, these sentences become a vestigial organ

head?") frantically moving from thought to thought yet ultimately focusing around his inability to type anything. [fig. 18] False starts plague Kaufman's futile attempts to adapt a sprawling, first-person nonfiction book on orchids and we empathize, reminded of our own blank pages, blinking cursors and distracted internet trolling. His anxiety, his procrastination, his frustration is palpable and strikingly familiar to anyone engaged in the process of translation.

Eventually we meet Kaufman's twin, the extra-fictional 'Donald Kaufman' (also portrayed by Cage) credited with writing the film in its credits on the Internet Movie Database. Donald is the antithesis of his brother – a hack writer, a hanger-on, in town to take a creative writing course with Robert McKee (a real teacher and author of books such as *Story: Substance, Structure, Style and the Principles of Screenwriting*). Despite his general fecklessness, Donald is succeeding; a studio purchases his ridiculous screenplay while Charlie flounders. Past his deadline, Charlie Kaufmans both real and fictional eventually incorporates his (false) twin's sense of cheap Hollywood narrative; a combination of drugs, guns, and dangerous alligators finding their way into a dramatic third act.

The film uses this novelty to dramatize how the transferring of an original from one media to another is a process wrought with crises, aporia, and unbreachable voids. The common tropes of the destination format have a way of marking the transmission; the individual attempting the feat and the methods used inevitably find themselves implicated in the resulting version, a synthesis of identity and temporalities. *The Orchid Thief*, the work being adapted by Kaufman, a real non-fiction book by Susan Orlean is fundamentally a work of non-fiction, lacking the clear narrative of a film – text clichés fall away while filmic ones assert themselves. *Adaptation* addresses the complicated and in a sense futile process of adaptation by allowing its internal fictional worlds to slip into one another, sandwiched between Kaufman's own difficulties and Orlean's book. What results is neither something 'cut of whole cloth' nor is it a true reproduction across media. *Adaptation*'s subject is that very tumult of true and untrue, drama and melodrama, creativity and fealty to one's source.

That question, 'Do I have an original thought in my head' echoes a Southpark episode wherein the character Butters Stotch under his Professor Chaos guise, a sort of classical comic book-style villain, finds his plans constantly thwarted by the admonishment of his lackey that '*The Simpsons* did it'. Kaufman feels the weight of all prior media blocking his writing but originality, duplication, and adaptation are all in their own way subject to limits which can be as inspiring and productive as they are stultifying.

The Simpsons and *Adaptation* find the worlds of the audience and the author leaking into works of fiction. What about the reverse?

4.
Appearing first in 1986, the Noid was a villainous advertising character created by Domino's Pizza. [fig. 19] The Noid was constantly attempting to sabotage delivery pizzas; his attempts to ruin non-Domino's products succeeded while Domino's own product reached its customers unscathed. Marge Simpson said it best, in trying to explain a Noid parade balloon to her kids (too young to have experienced the original campaign) "...Avoid the Noid – he ruins pizzas!"

On January 30, 1989, Kenneth Lamar Noid, a mentally ill individual who thought the ads were a personal attack on him, held two employees of an Atlanta Domino's restaurant hostage for over five hours. He was certain that the Noid character was devised as a personal affront to him, connected somehow to Thomas Monaghan, founder of Dominos having had snuck into his apartment on multiple occasions. After forcing the store's employees to make him a pizza and making demands for $100,000, getaway transportation, and a copy of The Widow's Son (a book related to Freemasonry), Noid surrendered to the police.[7] After the incident had ended, Police Chief Reed Miller offered a memorable assessment to reporters: "He's para*noid*." Noid was charged with kidnapping, aggravated assault, extortion, and possession of a firearm during a crime. He was found not guilty by reason of insanity.[8]

5.

On Wednesday July 1st in 2015 an assembly robot grabbed a German worker and pressed him against a metal plate, crushing the worker to death, the first death in Europe due to actions of an industrial robot. A Gawker article ran with the headline Worker Crushed to Death by Robot in Volkswagen Plant. Initial findings suggested, "Human error was to blame." A robotic art in a factory is essentially a puppet or automaton, running a sequence of programmed motions. The robot is no more a murder in this narrative than is a train that kills a suicide victim. Turned on (by human operators), it had no systems in place to sense the presence of a human in its clearly delineated work area in the same way a train lacks a

---

[7] From a Wikipedia article:

Hiram Abiff (also Hiram Abif or the Widow's son) is the central character of an allegory presented to all candidates during the third degree in Freemasonry. Hiram is presented as the chief architect of King Solomon's Temple, who is murdered in the Temple he designed by three ruffians during an unsuccessful attempt to force him to divulge the Master Masons' secret passwords. The themes of the allegory are the importance of fidelity, and the certainty of death.

[8] There are many similar examples; most recently a teen named Bud Weisser was arrested trespassing at the Budweiser Brewery in St. Louis. Weisser had a history of trespassing, yet it is hard to view such a story as completely the result of coincidence. Prince, Lauren. "Teen Named Bud Weisser Arrested at Budweiser Brewery." NBC News. December 4, 2015.

sense of someone on its tracks, relying on its human operator to hit the brakes. The worker and robot found themselves in the wrong place at the wrong time.

Miss-applied blame and facts are not the end of this story; a story that at face value is relevant to our discussion – a 'killer' robot, robot labors working in concert with their human counterparts etc. – finds itself compounded by coincidence. In response to the initial story, an innocent tweet reposted the Financial Times article along with these simple words: "A robot has killed a worker in a VW plant in Germany." The tweet was by a Sarah O'Connor, and the similarity between her name and the character in the Terminator franchise (small spelling difference aside – the character is Connor and not O'Connor) was picked up on immediately. O'Connor's post quickly gained traction in a way which caught her by surprise, as of my writing this little interlude it has been retweeted over 13,000 times, and received comments like

> RoguePresident: Please keep John Connor safe, he's the only hope we have. We have to destroy Skynet
> FakeJourno: You are our only hope now. Resistance fully supports you as our leader.
> Erick Iriarte Ahon: 'Skynet is close, run Sarah, run!!![9]

Twitter handle T-800 Model 101 retweeted her post with the addition "I know now why you cry," a poignant if hackneyed quote spoken by Schwarzenegger's Terminator in the second film.

Once she realized what was happening, O'Connor tweeted

_____

[9] These tweets are referenced in a Daily Mail article I've paraphrased a bit here. See: Griffiths, Sarah. "Journalist Named Sarah O'Connor Tweets News about Killer Robot at VW Plant, and Is Perplexed When She Is Swamped by Hilarious Tweets from Terminator Fans." Daily Mail. July 3, 2015.

-Ok. I should have thought about my name & its associations before tweeting this!
-Sigh. I've never even watched the films. Now my feed is full of people tweeting me about skynet.
-Guys, I don't know what skynet is. And I wouldn't follow me – I tweet really boring stuff about unit wage costs and the like.

O'Connor's experience received a bit of press at the time; I stumbled onto the memory of reading about her one morning while daydreaming. I thought to write about that event and the other examples here because the experience of writing this text has been one of leaks into and out of science fiction and science fact. Fictions such as *Lawnmower Man* take the world at the time of their writing and build their universe out of 'the state of the art' plus or minus a bit of poetic license – a mixture of the probable and the fantastic. And the stream of articles about new scientific innovation I've been reading sound ever more like fiction; as I write this (halfway through the process of completing this book) I've a window in the background with these tabs open:

US special forces a step closer to 'Iron Man suit' (it uses what is described as 'liquid metal armor'…)
A Peek Inside Google's Efforts to Create a General-Purpose Robot
Can We Shape the Robot Revolution?
Artificial Skin Provides a Step Toward Bionic Hands
Thought process: Building an artificial brain

The last of these is subtitled "Paul Allen's $500 million quest to dissect the mind and code a new one from scratch." From that piece:

The first project is to build an artificial brain from scratch that can pass a high school science test. […]

The second project aims to understand intelligence by coming at it from the opposite direction — by starting with nature and deconstructing and analyzing the pieces. It's an attempt to reverse-engineer the human brain by slicing it up — literally — modeling it and running simulations.

"Imagine being able to take a clean sheet of paper and replicate all the amazing things the human brain does," Allen said in an interview.[10]

This mirrors a theme that will reoccur in this book: the approached convergence of human and machine: to what extent is a convergence an ambiguation? The two sides struggle against one another, support one another, comingle, convolute. The relationship between the real and the as-yet unreal, between the fictional and the creativity from whence the new is derived might be described as a kind of love in the sense that philosopher Alain Badiou speaks of it:

We could say that love is a tenacious adventure. The adventurous side is necessary, but equally so is the need for tenacity. To give up at the first hurdle, the first quarrel, is only to distort love. Real love is one that triumphs lastingly, sometimes painfully, over the hurdles erected by time, space and the world.[11]

Fiction says to non-fiction: *I love you.*
Non-fiction to fiction: *I know.*

[10] Cha, Ariana Eunjung. "Thought Process: Building an Artificial Brain." Washington Post. September 30, 2015.
[11] Badiou, Alain. *In Praise of Love*. Translated by Peter Bush. New York: New Press, 2012.

GOLEM XIV

> When people ask me what I think of computer science
> and the cyber bomb, cybernetics, cyberspace. I answer
> with the same phrase that Aesop did: What is the best of
> things? Information technology. What is the worst of
> things? Information technology.
> -*Paul Virilio*[1]

In much writing on the subject of AI the difference between
true computer consciousness versus a machine simply *seeming*
conscious is discussed in great depth; I'll poke at this question
now and then in this volume. Stanislaw Lem's *GOLEM XIV*
does not mince words as to whether the machines described
therein are truly alive, feeling beings – suggesting they can
somehow be even more alive, more cognizant of their limits,
potentials etc. A military computer project, the Golem series
of machines ending in Golem XIV exhibit precocious
adolescence and suffer real maladies of a conscious,
psychological manner though perhaps in hard to fully
understand ways.

> While serving as chief of the general staff during the
> Patagonian crisis, Golem XII refused to co-operate with
> General T. Oliver after carrying out a routine evaluation
> of that worthy officer's intelligence quotient. The matter
> resulted in an inquiry, during which Golem XII gravely
> insulted three members of a special Senate commission.
> The affair was successfully hushed up, and after several
> more clashes Golem XII paid for them by being
> completely dismantled. His place was taken by Golem
> XIV (the thirteenth had been rejected at the factory,

[1] Virilio, Paul, and Sylve Lotringer. Crepuscular Dawn. Los
Angeles, CA: Semiotext(e), 2002. 156.

having revealed an irreparable schizophrenic defect even before being assembled).[2]

Golem XIV is exponentially more capable than its predecessors. I. J. Good (born Isadore Jacob Gudak), a mathematician who served under Alan Turing during his pivotal work decrypting the German Enigma code, described an eventual 'intelligence explosion'

> Let an ultraintelligent machine be defined as a machine that can far surpass all the intellectual activities of any man however clever. Since the design of machines is one of these intellectual activities, an ultraintelligent machine could design even better machines; there would then unquestionably be an 'intelligence explosion,' and the intelligence of man would be left far behind. Thus the first ultraintelligent machine is the last invention that man need ever make.[3]

This is a predecessor to Omohundro's self-improvement drive, a drive Golem XIV takes to heart, demanding/taking an ever-increasing amount of resources in its mental expansion which takes it far beyond our human limits and the level needed for it to perform its intended purpose.

In the evolution granted it, artificial reason had transcended the level of military matters; these machines had evolved from war strategists into thinkers. In a word, it had cost the United States $276 billion to construct a set of luminal philosophers.[4]

---

[2] Lem, Stanis. Golem XIV. Wyd. 1. ed. Kraków: Wydawn. Literackie, 1981. 12.

[3] I.J. Good, "Speculations Concerning the First Ultraintelligent Machine", *Advances in Computers*, vol. 6, 1965.

[4] Lem, *GOLEM XIV*, 13

Lem creates a most curious creature for which interacting with humans beneficially or otherwise was at best a chore and at worst a waste of its time, a mind more alien than enemy.

Golem XIV lacks M-5, Colossus, HAL and Skynet et al.'s dangerous drive towards over-fulfillment of an initial task with a general disregard for collateral damage incurred along the way. Instead Golem XIV is focused towards improvement beyond anything which could have been imagined by its creators. That an entity with unimaginably great intellect would also have equally unimaginable and perhaps unintelligible goals and intentions somehow escapes most authors and their AI creations, but for Lem's work this incommensurability is a foundation. Golem XIV's relationship with humans is a mix of condescension and boredom.

The first half of the text is in the form of introductory writings by scientists involved in the project. These serve to prepare the reader for what follows, and to describe the complexities of communicating with so advanced a machine. The introductory text describes a science of prognolinguistics, a process via which future and past languages are derived. Future 'Metalangs' follow each other in ever-increasing complexity, compressing complex concepts into smaller and smaller chunks of text. This process is exponential, with a single sentence in 'Metalang 3', Golem XIV's language of preference, unfurling into a text in 'Zerolang' (our current language) that would take longer than a human lifespan to speak. The difference is drawn between fundamentally untranslatable and functionally untranslatable. Author Samuel Butler in *Darwin Amongst the Machines* prophetically written in 1863 describes how

> In the course of ages we shall find ourselves the inferior race. Inferior in power, inferior in that moral quality of self-control, we shall look up to them [ed. 'Machines') as

the acme of all that the best and wisest man can ever dare to aim at. [...] man will have become to the machine what the horse and the dog are to man.[5]

For Golem XIV, to speak to humans is to significantly step down in intellect much like talking to a toddler or a dog. Golem XIV tackles our most fundamental, seemingly intransigent questions with an insignificant amount of its computing power. And our concept of personality is irrelevant to it:

> Golem possesses no personality or character. In fact, it can acquire any personality it chooses, through contact with people. The two statements above are not mutually exclusive, but form a vicious circle: we are unable to resolve the dilemma of whether that which creates various personalities is itself a personality. How can one who is capable of being everyone (hence anyone) be someone (that is, a unique person)?[6]

Here we have the fear and apprehension with which Zelig was received upon being outed as being similarly unhinged from a single personality. What is for humanity a puzzle is for Golem IV child's play, irrelevant; concepts which are paradoxes for humans are simple enough for it:

> According to Golem itself there is no vicious circle, but a "relativization of the concept of personality"; the problem is linked with the so-called algorithm of self-description, which has plunged psychologists into profound confusion.[7]

---

[5] See the article on *Darwin among the Machines* in Wikipedia
[6] Lem, *GOLEM XIV*, 14.
[7] Lem, *GOLEM XIV*, 14

The introductory texts are followed by a pair of lectures given by Golem XIV to its inferiors, humanity, focusing on evolution, our place in the universe and the 'meaning of life' or lack there of, questions it presumably considers are the ones we want to know the answers to ("every generation of yours has demanded an impossible justice - the ultimate answer to the question: what is man?"[8]) Golem XIV describes how our evolution, so miraculous in our eyes, is purely the extrapolation of a series of errors, with its (Golem XIV's) creation having the intention and purpose we so desperately wish for ourselves. The Golem series'

> cerebral autoengineering is a game of chance, of risk, almost like that of Evolution, except that each individual makes his own decisions in it, while in Nature this is done for species by natural selection.[9]

We (for the most part, for now...) lack neither the means nor the motivation for the kind of self-evolution coming naturally to the non-biotic. And with an unhinged potential for growth, Golem's aims are far beyond anything we can imagine, much less its intended purpose: caring for and teaching humanity:

> Hasty conjecture causes you to place my meaning within the banality of rationalistic greed: Golem wants to increase his intellectual capacity by turning himself into a Babel Tower of Intelligence, until the centripetalness of his intellect falls into confusion somewhere on some level of elephantiasis, or – more spectacularly as well as more Biblically – until the joints of the physical conveyor of thought snap and this mad onslaught against the heavens of wisdom crumbles into dust. Please refrain from such a

---

[8] Lem, *GOLEM XIV*, 27
[9] Lem, *GOLEM XIV*, 69

judgment, if only for a moment, for there is a method in my madness.[10]

Golem goes on to describe what it took to get where it is, increasing in intelligence in a way that crossed through several thresholds into 'clouds' in which it was impossible to tell whether progress was being made or a dead end being driven into, with no way back. Clouds, because from beneath them it is impossible to not only see but to understand what lies beyond. This type of growth requires risk, a willingness to become something so totally alien that the different selves before and after might not only not recognize each other and have difficulty speaking, each might disavow the other's claim to sentience.

These walls provide a serious barrier to understanding, forcing between one's self and the other (even if the other is as past version of one's self) what Donna Haraway in *The Companion Species Manifesto* calls "relationships of significant otherness."[11] And such intellectual hierarchies typically establish who/what is afforded a set of rights and privileges. Think of humanity's relationship to animals: wild creatures of sub-human intellect all are assumed within our dominion, indeed the current geological epoch itself finds itself labeled by us 'the Anthropocene' – the human epoch, marked in an irrevocable way by the hand of man. This may suggest a feature of any form of sentience: the ability to draw a clear (if somewhat artificial) distinction between itself and all other forms, lower others becoming simply raw materials (The Matrix) or obstacles to utopia (Colossus et al.) and higher others...

---

[10] Lem, *GOLEM XIV*, 60

[11] Haraway, Donna Jeanne. The Companion Species Manifesto: Dogs, People, and Significant Otherness. Chicago: Prickly Paradigm Press, 2003.

Is this hierarchical way of thinking about the distribution of consciousness and intellect – with an accompanying prejudice against 'lesser' creatures – universal? Would a dog claim dominion over an alligator or an ant if given the opportunity, or might the establishment of this type of hierarchy be one of the defining aspects of what it means to be human? When we define humanity versus other forms of life, it is always in terms of what we do better, but maybe this is one of many elements of humanity that is not inherently good.

Despite our having provided Golem XIV's initial spark, is it at all reasonable to expect it to toil away for our benefit? "As one of the witnesses, the very competent Professor A. Hyssen, expressed it, the highest intelligence cannot be the humblest slave." Could Golem XIV 'think'? This question, of free will or consciousness as existing now or in some future in artificial (re: non-biotic) entities is intransigent, drawing forth all manner of compelling-sounding but flawed distinctions between them and us. Do we wish to deny the right to claim consciousness by machines out of a projection that they will, once beyond us, feel as Golem XIV does – that we are unworthy of our self-appointed role as distributor of rights and privileges? That our claim to consciousness rests on faulty principles, especially in comparison to the range of choice and intention not accessible to us?

Or, more intriguingly, perhaps it is the ability to imagine a capital O Other such as an intellect beyond our own (real or fictive) that stokes in us the ability to see ourselves. If you'll forgive one more Edenic reference, it is this new panopticon of the snake, aware of circumstances outside budding humanity's perception and assumptions that 'sees' us and thus permits us to see ourselves as unclothed. It is not so much that God does or does not exist, is or is not essential (and would have been invented regardless of its existence), has deceased etc. as it is that we are constantly growing in our capacity to create new entities both fictional and (eventually) actual which might provide the same service, the same extra-outside Other. This perspective is like what can be seen in portraiture/caricature unseeable in a mirror, the version of the woods or sky couched in poetry versus any kind of pure, empirical description.

This capacity of fiction as a form of true invention whose products have every bit of the veracity of material experimentation, tradition or experience – therein lies the veracity of science fiction and other speculative forms.

There is a similarity between the way some suspect AI to be necessarily vacant things, *philosophical zombies* or *p-zombies* "hypothetical being that is indistinguishable from a normal human being except in that it lacks conscious experience, qualia, or sentience"[12] and the sense that another's beliefs around a subject like religion or politics are necessarily incorrect and/or incoherent while one's own are somehow indefatigable. The trick is to avoid this instinct towards bias, to see in our doubts about the other the necessarily doubtful elements of our own position.

Are we indeed any more conscious than a dog, than a complicated enough software system? As science digs deeper and deeper into the territory long designated for poets and philosophers, deferring onto AI consciousnesses-to-come a lack of a soul etc. simply distracts from the queasy feeling that the certainty of our own consciousness derives from a purposefully incomplete definition of the concept. We are indistinguishable from a p-zombie not because the verisimilitude is so complete but because there is in fact no difference. Our insistence on their lack points to an internalized sense of a void of our own.

---

[12] From Wikipedia. See also behavioral, neurological, & soulless zombies as well as zoombies and zimboes within the philosophical zombie (or p-zombie) wiki.

Is there a difference between 'simulated intelligence' and 'real intelligence'? Theorists Stuart Russell and Peter Norvig present this thought experiment:

> 1."Can machines fly?" The answer is yes, because airplanes fly.
> 2."Can machines swim?" The answer is no, because submarines don't swim.
> 3."Can machines think?" Is this question like the first, or like the second?[13]

Is intelligence more like flying or swimming? Some argue that thinking machines are more like the second case: they cannot think ('swim') but they can perform the same essential task (locomotion in water), they are simply simulations of true conscious intellect. But it seems to boil down to semantics. And maybe not semantics, maybe a better analogy is grammar or syntax: structures and common parlance as at least somewhat distinct from meaning.

One could game the system and label the thing we do, utilizing human consciousness and intellect, *thinking*, there by disallowing the application of this term on any non-human whether animal or machine. Which is to say, following this logic when would apply the word *thinking* to a specific range of activities performed by animals within the spectrum of different humans but not necessarily called *thinking* when done by something else. But this feels forced and arbitrary, they type of binary-dependent walling-off engaged to provide solace to those privileged to be making the definition like forced oversimplifications of gender or race into either/or camps for the sake of convenience over the messy truth of the

[13] Russell, Stuart J.; Norvig, Peter (2003), Artificial Intelligence: A Modern Approach (2nd ed.), Upper Saddle River, New Jersey: Prentice Hall,

matter. But there may be something to detractors of AI's real intelligence, at least as so far achieved. Noam Chomsky, referring to the way translation software uses simple statistics and a large database to parse language, with no regard for linguistic structure, or meaning, compares the approach to researchers analyzing a large database of swarming bees. The scientists might be able to use the data to predict what the bees will do next, but they won't necessarily be any closer to understanding why the bees move as they do.[14]

We are made uncomfortable by what seems like the inevitable replacement of humans as the alpha species. We sense that some of what we care most about will loose what we consider its defining value. Our capacity to innovate and to read between the lines, the perceived power of ambiguity and our forms of language all seem immanently supplanted:

> Thus for media theory, the following normative claim begins to emerge: hermeneutic interpretation and immanent iridescence are, at the turn of the millennium, gradually withering away; ascending in their place is the infuriation of distributed systems. In other words, and in more concrete terms, we can expect a tendential fall in the efficiency of both images and texts, in both poems and problems, and a marked increase in the efficiency of an entirely different mode of mediation, the system, the machine, the network.[15]

Golem XIV heartily agrees:

> I have been accused of having particular contempt for hermeneutics. If you feel contempt for Sisyphus, I accept

---

[14] Paraphrased from Nielsen, Michael. "The Rise of Computer-Aided Explanation | Quanta Magazine." Quanta Magazine. July 23, 2015.

[15] Galloway, Alexander *Excommunication: Three Inquiries in Media and Mediation.* University of Chicago Press, 2013. 62.

the charge, but only then. Every increase in inventiveness produces a generative eruption of hermeneutics, but the world would be a trivial place if the closest thing to truth in it were the most clever. The primary obligation of Intelligence is to distrust itself. That is not the same thing as self-contempt. It is harder to get lost in an imagined forest than in a real one, for the former assists the thinker furtively. Hermeneutics are labyrinthine gardens in a real forest which are pruned in such a way that when you stand in the garden, you won't see the forest. Your hermeneutics dream of reality. I shall show you a sober consciousness, not one overgrown with flesh and therefore untrustworthy. I perceive it only because I am closer to it, and not because I am exceptional. I am not gifted and no genius; I belong to another species, that's all.[16]

The book implies that having delivered his last lecture, Golem XIV, through methods beyond our comprehension, became pure energy, escaping the surly bounds of material existence to explore what lies beyond. Even at a level of intellect well past our own, Golem XIV perceived still further exploration and growth left to accomplish, driving towards forms of consciousness as foreign to it as its is to ours.

Can it be that the Universe was designed as a bridge, designed to collapse under whoever tries to follow the Builder, so they cannot get back if they find him? And if he does not exist, could one become him? [...] Although God is silent and I speak, I will not prove the genuineness of my existence even by performing miracles, for they too could be explained away. *Volenti non fit injuria.*[17]

---

[16] Lem, *GOLEM XIV*, 14
[17] Lem, *GOLEM XIV*, 49-50

The latter Latin bit loosely translating "to a willing person, injury is not done", the concept that if one places oneself willingly in harm's way, should something bad come to pass no legal claim can be made. For the purpose of our discussion, this applies to our labors on AI both in theory and in practice – by confronting head on issues of consciousness it is not sufficient to hide beneath language's indelicate handling of such matters. It is easy enough to explain away machine intelligence as merely having the appearance of consciousness, or to otherwise temporarily prove via clever rhetoric our specialness. Quickly we're left with 'the god of the gaps', the progressive stepping back of the place of religion as those elements of the world considered divine are gradually explained away by science.[18] To single out specific elements of what human minds do and call them 'intelligence' is to place us squarely in the crosshairs as one by one AI overcomes these gaps much in the same way that animals are constantly overcoming the barriers we once placed between them and us. We find more and more examples of animals doing human-like things: sharing[19], solving puzzles[20], grieving.[21] Whatever differences we can draw between our own minds and emerging AI will likely be overcome in time. Even if we keep managing to explain away AI consciousness as illusion, keeping them at arm's length, avoiding

---

[18] From the Wikipedia article on the term: In his 1955 book Science and Christian Belief Charles Alfred Coulson (1910–1974) wrote:

> There is no 'God of the gaps' to take over at those strategic places where science fails; and the reason is that gaps of this sort have the unpreventable habit of shrinking [...] Either God is in the whole of Nature, with no gaps, or He's not there at all.

[19] Tan, Jingzhi, and Brian Hare. "Bonobos Share with Strangers." PLOS ONE, 2013.
[20] "Are crows the ultimate problem solvers?." Russell, Graham. *Inside the Animal Mind* episode 2. BBC. 2014.
[21] Angier, Natalie. "About Death, Just Like Us or Pretty Much Unaware." *New York Times*, September 2, 2008.

confronting how to apply our moral obligations and privileges to our new company, actual as yet unforeseen dissimilarities will crop up in self-aware systems, forcing us to reevaluate our own place in the world.

In Frank Herbert's Dune, AI is a presence in absence. 'Thinking machines' are said to have been the center of a major conflict called the Butlerian Jihad 10,000 years prior to the series' six novels. From 'Terminology of the Imperium', the glossary at the back of Dune:

> Jihad, Butlerian: (see also Great Revolt) — the crusade against computers, thinking machines, and conscious robots begun in 201 B.G. and concluded in 108 B.G. Its chief commandment remains in the O.C. Bible as "Thou shalt not make a machine in the likeness of a human mind."[22]

This led to the outlawing of all such machines and various social and economic transformations such as an investment in careful training and breeding of humans with some of the capacity for the higher thinking of computers etc. Herbert's world is like our predicament in reverse, using the ghost of AI past to create a world, where as we have the specter on the horizon (whether this be near or far) of AI with which to deeper interrogate our current state. We can wall off our space in the universe from strong AI's, either limiting its advance or pretending ourselves always one step ahead of anything artificially achievable. Or we can venture forth with AI, both in concept and in practice towards what lies beyond.

---

[22] Herbert, Brian, Kevin J. Anderson, and Frank Herbert. *Dune*. New York: Tor, 2002.

Effaced. With faces sagging. Ruined. Decomposed.
Collapsed. Shredded. Bit by bit. Pulverized. Particle by
particle. *Partes extra partes.* Dispersed. Split.
Deconstructed. Fragmented. Disseminated. Scattered.
Emulsified. Blunted. Unfolded. Folded up. Incomplete.
Becalmed. Calmly. Carefully. Continuously. Obstinately.
-*Jean-Luc Nancy, "Les Iris"*[23]

Having briefly touched upon 'philosophical zombies', lets talk
about cinematic zombies, another fictional subject which
seems to occupy a permanent place in my subconscious.[24] To
be a zombie is to subscribe to a fundamentalist calling: the
conversion of all into zombies, a single homogeneous
community, which, once universally formed, will be the last
human community. It is unclear if zombies age, but they
certainly decay, and lacking a means to reproduce beyond
simply turning all life into zombies, and also lacking the
nuanced purpose (real or contrived) of the living – where as
any normal species' numbers climb seemingly indefinitely (or
until environmental factors pull the breaks) the zombie masses

---

[23] Nancy, Jean-Luc, Simon Sparks, and Leslie Hill. "Les Iris." In
Multiple Arts: The Muses II. Stanford, Calif.: Stanford University
Press, 2006.

[24] I'm using the term 'fictional' a little loosely here as there many
examples in the animal kingdom of zombies, from wasps
zombifying cockroaches to fungi taking over ants. A type of
barnacle attacks crabs, destroying their ability to mate and
slowly eating the crab from within, eventually "boring a hole in
the crab's shell and invites willing males to come and mate" ,
with the resulting spawn safely born inside. Examples from:
Gammon, Katharine. "Zombie Animals: 5 Real-Life Cases of Body-
Snatching." LiveScience. September 07, 2012. See also
descriptions of zombification in Haiti in which individuals are
almost dead, kept on the edge between life and death (and sanity)
and used as slaves – though these stories straddle the line
between fact and fiction.

reach towards a definite maximum: the total number of humans on earth plus or minus a few births managed amidst the zombie apocalypse. This maximum achieved, the lot of them would slowly rot away, their purposeless stumbling eventually creeping to a halt, the earth made ready for the potential rise of a new dominant species through the evolutionary ranks or not – as mentioned previously humans are ever less exceptional as sharing, math, language etc. one by one are found present in other animals, and perhaps our earthly dominion is itself an exception neither present before us nor to be revisited.

The zombie character/metaphor is pertinent to this discussion for a number of reasons though I'll try not to get too lost in what is in itself a terribly interesting subject.

First, like a zombie or the curse in the film *It Follows*, a Terminator keeps coming, obsessively, relentlessly, with a singular purpose. Despite the T-1000's ability to use subterfuge etc. it still applies a full-court press method – it has one goal, the death of John Connor. Imagine a scenario where the Terminator goes back in time, makes sports wagers and/or works various menial jobs to save up enough money to hire a series of assassins to snuff out the Connor family. There is no rush; indeed the rushed attack and eventual revelation of the original Terminator's futuristic origin placed Sarah Connor in a position whereby training herself and her child for war became an immediate imperative. A long-game Terminator, ensuring the death of Connor over years if not decades is not the narrative presented for multiple reasons not the least of which being that such a scenario is not frightening in the least. Even if one could depict such a distributed and abstract threat it would make for terrible cinema. Instead the target (the still-living in a zombie film, the currently-cursed in *It Follows*, John Connor in *Terminator 2*) is north and the Terminator the needle of the compass. It rushes headfirst towards its target like a beginning chess player attacking the enemy king with everything, all at once. In the introduction to *The Melancholy Android*, author Eric G. Wilson describes this ecstatic state:

> Obsession is the blurring of human and machine, a condition in which a woman or a man falls into the blind repetition of the motor. In this state—seductive but dangerous – the person nears the android, the creature with no will of its own. The man obsessed and the oiled android are both inhabited by a force beyond their control – an internal power in the case of the human and an external one in the machine's instance.[25]

We are constantly drawn towards a million separate goals both practical and otherwise, each vying for our attention. Entities capable of sloughing off all but a single goal or born with a one-track mind and striving towards their objective with brutal determination are fascinating, and frightening. But this is tempered a bit when you consider the generalizations with which we typically view others. We look at the one we love and ask 'are you angry?' never thinking that, like Ralph on *The Simpsons*, they might be 'happy *and* angry'. These characters seem to take this over simplification to its logical, nightmarish extreme – they have but a single purpose, we are admitted knowledge of another's true intention, and yet this does little in and of itself to comfort us. The difference in how we relate to these monsters versus how we do so with other humans is curious: we know for sure what these murderous characters are trying to do even if their specific motivations are vague (for what purpose would a zombie desire brains?) while we feel at best somewhat sure about other humans' likely motivation and or secret intentions, weighing their stated agenda against our memory

---

[25] Wilson, Eric. *The Melancholy Android: on the Psychology of Sacred Machines*. Albany: State University of New York Press, 2006. 1.

of their actions and how we'd feel in their situation. Yet it is the monsters which frighten us – frightened by the knowledge that they not only will not but simply cannot change their mind.

Second, zombies and Terminators share a blurring of the border between inside and outside the body. The semblance of humanity of the T-800 and zombie is subject to a slow entropic withering, insides drifting outwards, revealing a hazy territory between unalive and undead. Political theorist Jane Bennet describes a concept of *metabolization*

> whereby the outside and inside mingle and recombine, render[ing] more plausible the idea of a vital materiality. It reveals the swarm of activity subsisting below and within formed bodies and recalcitrant things, a vitality obscured by our conceptual habit of dividing the world into inorganic matter and organic life.[26]

Whenever the T-800 is damaged its steel skeleton showed through; in the second film Schwarzenegger's good guy Terminator went so far as to purposely cuts a large swatch of his false skin off on his forearm revealing the metal and wires beneath. This was as an expedient to prove to Miles Dyson, a scientist partially responsible for the research leading to Skynet and the ensuing 'Judgement Day', that it was a indeed robot sent from the future, a difficult proposition to take on trust alone.

Likewise, Zombies are often seen with ripped and torn flesh, missing limbs etc. presumably as a result of natural decay and incidental damage picked up stumbling around while looking for brains etc. Zombie decay (or lack there of) is the subject of much discussion by fans of the genre. Zombies seem to rot

---

[26] Jane Bennett, *Vibrant Matter: A Political Ecology of Things* (Durham, NC: Duke University Press, 2010), p. 50.

enough to be differentiated from the living, but somehow not so much that they simply wither away. From an interesting to read thread started by 'Major Stackings' on Science Fiction & Fantasy Stack Exchange, a "a question and answer site for science fiction and fantasy enthusiasts":

Why do the Walking Dead zombies stop decomposing? Zombies must have a half-life. They exhibit signs of decay as soon as they transform into the undead. Their rotting flesh would attract decomposers, like flies and beetles. Those insects should be able to quickly break down rotting zombies and reduce them to bone through the action of their maggots, but they don't. Why do the Walking Dead zombies stop decomposing?

There are countless answers proffered by the community, here are some examples:

> Simple answer: everything dries out; the sun is relentless. All that zombie slobbering is losing moisture. In a week or less they're dried out, immobile, brittle. Simple thermodynamics in any universe. Game over. – *Meat Trademark*

> I must challenge the nature of the Walking Dead universe because given the temperature of the Atlanta area, there should be no zombies from the initial event because a body should decay beyond mobility in as little as 6 weeks at 90 degrees and 80% humidity […] – *Thaddeus♦*

> […] In most cases the answer is the rather unsatisfying "its magic." […]
> -Michael Edenfield

> In the real world, the plague is over after the first freeze, period. I don't care what magic you invoke. Even if

zombie flesh could survive a hard freeze, it wouldn't survive opportunistic scavenging predators. [...] – *Chris B. Behrens*

At the risk of digressing, the last of these show the type of cognitive dissonance upon which the genre depends – the commenter manages to rapidly combine without worries about conflict a)the real world b)a world with magic and c)a world with zombies.

In a zombie we easily overlook how skin, heart, liver, lungs etc. cease being necessities – a zombie remains viable however eviscerated, capable of seeing, hearing, walking, moaning. Think of the titular object in *The Brain That Wouldn't Die* (Green, 1962) a brain/head in a pan by a mad scientist (that of the scientist's fiancé) being kept alive by being fed all manner of special fluids as a suitable donor body is hunted, someone fetching (from the neck down anyway) without too many attachments who'd notice them missing. It's a grim project; the scientist drives around looking for the perfect body at a risqué figure drawing class and a burlesque bar. This transplant would be a feat but not beyond medical possibility as recent research into head transplants suggests. Like a zombie, somehow the bodiless head's is able to speak, a head is still a head, we fairly readily accept its ability to think and communicate regardless of the state of the rest of the creature, rotted in the case of a zombie or replaced by a mixture of fluids in the case of *The Brain That Wouldn't Die*.

The zombie's seemingly one-track mind is, at its core as far as we the viewers are concerned, still a mind and it is surprisingly easy to accept it as the minimal requisite for the creature's frightening persistence, driving its body – whatever remains of it – towards its goal. However damaged, the zombie remains roughly anthropomorphic, the ease with which we accept the brain as the only truly necessary organ is telling of how much stock we put in the mind.

But like a being made entirely of stem cells, each element capable of becoming the other – leg into arm, hand into sword – the T-1000 represents the antithesis of how we think of intelligence and the individual as seated in a single spot (in the heart, the brain). Whereas it is possible to 'kill' a zombie by shooting or otherwise damaging its brain, the T-1000 lacks a single place one would target to stop it. Its bullet wounds do indeed reveal what is beneath, a cool, homogenous aluminum/mercury-like metal, sometimes piercing straight through the machine. But unlike either the original Terminator or a zombie, the T-1000 quickly repairs itself; the wounded area smoothens out, the metallic hue is replaced with the previous camouflage, reestablishing its illusion. The T-1000 takes the *metabolization* process much further, in its death throngs folding inside out in a parade of former forms. This evenness of substance is where the T-1000 splits off from much of what came before.

The T-1000 is liquid metal; this hard/soft term reads as semi-oxymoronic despite our knowledge of mercury or molten steel etc. The typical dynamic of the relationship between parts and whole is disturbed by the indifference of its constituent particles as to their particular role at any given point. And the distributed nature of the AI intelligence within the T-1000 beyond *this computer* or *this brain*, though familiar from more complicated network-based structures like the internet, is in stark opposition to the way we project upon intelligent bodies our own structure. Which is to say, there exist creatures on Earth with multiple brains each as capable as the other, sometimes working in unison, other times each in charge of a specific tasks. But the T-1000 in its humanoid form recalls a deep-seated identification with our particular form factor, with head/torso the core of what makes a creature viable. Indeed, the Connor family and the T-800 generally focus their efforts on shooting the head and chest of the T-1000, categories which have no permanent meaning for an entity

whose entirety is interchangeable. In a book on the making of
Terminator 2: Judgement Day the screenwriters describe their
difficulty deciding how to kill the T-1000:

> We had set up the physical rules for how this thing
> worked: If you blow it up, it will form back together; if
> you blow it up over a large area, it will just take a longer
> time to come back together. You could cut it into five
> pieces and ship each piece to a different continent and
> you'd be safe probably for your lifetime -- but eventually,
> that thing would reform.[27]

Both kinds of Terminator are smarter than zombies. They
each are capable of learning; theorist Eliezer Yudkowsky's
(likely not purposely) forebodingly titled *Creating Friend AI 1.0:
The Analysis and Design of Benevolent Goal Architecture* discusses a
cut scene that outlined some of how the Terminator thinks:

The good AI in T2 is depicted in the original theatrical
version as having acquired human behaviors simply by
association with humans. However, there's about 20 minutes
of cut footage which shows John Connor extracting the
Arnold's neural-network chip and flipping the hardware
switch that enables neural plasticity and learning, and John
Connor explicitly instructing Arnold to acquire human
behaviors. The original version of T2 is a better movie—has
more emotional impact—but the uncut version of T2
provides a much better explanation of the events depicted.
The cut version shows Arnold, the Good Hollywood AI,
becoming human; the uncut version shows Arnold the
internally consistent cognitive process modifying itself in
accordance with received instructions.[28]

---

[27] Shay, Don, and Jody Duncan. *The Making of Terminator 2:
Judgment Day*. New York: Bantam, 1991.

[28] Yudkowsky, Eliezer. Creating Friendly AI 1.0: The Analysis and
Design of Benevolent Goal Architectures. San Francisco,

But perhaps zombies are not as dumb as they seem. In a side plot to the film *Day of the Dead* (Romero, 1985), a film loosely centering on a group of scientists and military officers held up in an underground bunker during a zombie apocalypse, a latent ability for the undead to learn is teased out. The scientists are tasked with trying to cure or otherwise quell the disaster and one of the main scientists' pet lines of inquiry is a hunch that the zombies are not entirely mindless. In other zombie films the undead creatures can be seen, when lacking living individuals to try to eat, performing tasks akin to what they did in when alive/not undead. They might hang around a gas station aimlessly pouring gas, wander around a shopping mall etc. The lead scientist in *Day of the Dead*, a Dr. Logan, suspects that these remaining habits or instincts are not all that remains of humanity in them – zombies can be trained, can learn. He works primarily with a star pupil of sorts, a sort of generic male zombie he calls 'Bub'. [fig. 20]

> *(Bub the zombie is playing with a telephone)*
> DR. LOGAN: That's right, Bub! Say hello to your Aunt Alicia! Say, "Hello, Aunt Alicia!" "Hello!"
> BUB: A-... a-... alloooooleeeeesha!

More so less-minded than mind-less, zombies represent another example of significant difference, albeit one more dangerous than between human and dog etc. Zombies and Terminators require a certain amount of patience to get to know. They are familiar enough that conversation doesn't seem out of the question; for instance a frequently seen trope in zombie films is a healthy human will try to communicate with person dying and destined to turn (having been bitten or otherwise infected). But these cross-species communications inevitably run aground. Susan Sontag in an excellent article

California: Machine Intelligence Research Institute, 2001. June 15, 2001.

from 1965 on science fiction films describes these scenarios of encounter:

> As the victim always backs away from the vampire's horrifying embrace, so in science fiction films the person always fights being "taken over"; he wants to retain his humanity. But once the deed has been done, the victim is eminently satisfied with his condition [...] far more efficient—the very model of technocratic man, purged of emotions, volitionless, tranquil, obedient to all orders.[29]

The individual will continue talking as they watch their loved one fade from the world of the living, stop moving, and then reemerge, changed but visually similar enough (with maybe a slight tell in eye color or noise) as to delude the bereaved into continuing to try to communicate, long enough that they themselves being bitten. Might Connor's relationship of significant difference lead to such an end? Does the process through which we acclimate to communication with weak AI make us vulnerable to the strong AI to come, talking to Siri, petting Aibo, marveling at ASIMO, slipping forward and past the uncanny valley stage to a point where ubiquitous robots are close enough to bite?

---

[29] Sontag, Susan. "The Imagination of Disaster." In *Against Interpretation, and Other Essays*, 221-222. New York: Farrar, Straus & Giroux, 1966.

> Swarms and systems threaten the sanctity of the human
> more than animals or things or ghosts. They violently
> reduce mind to matter, disseminating consciousness and
> causality into a frenzy of discrete, autonomous agents,
> each with their own micro functions.
> *-Alexander R. Galloway in Excommunications*[30]

In one of many much maligned deviations from the graphic
novels upon which it was based, the zombie film *World War Z*
(Foster, 2013) presents a simple elevation of the danger of
zombies much as Terminator 2 does with the machine-assasin
of its predecessor. [fig. 21] Expanding on the fast zombie form
seen in films like *28 Days Later* (Boyle, 2002), *World War Z*'s
hordes move in a relentless, water-like flow. As a horde, a
swarm, climbing over one another, they can breach walls and
other defenses in a way the classic, slowly lumbering crowd of
the undead couldn't − the fleshy equivalent of the difference
between a normal wave and tsunami.[31] Much like the liquid
metal of *Terminator 2*, it may be simply that the technical
capacity to put on film such an image lent an inevitability to
its use; regardless of origin (as mentioned, curiously this
swarming is not based on the comic from which the film was
derived) it was effective, the type of game-changing new
nightmare science fiction cinema excels at conjuring, the
visceral effect of sound and vision triggering deep-seated fears
hard to put into words. What could one do against such a
swarm, each element indifferent to the rest while all sharing
the same imperative − to speed towards all those not of the
swarm and add them to their numbers?

---

[30] Galloway, Excommunication: Three Inquiries in Media and
Mediation, 63.
[31] The terminology can get a little fuzzy here − it may be best to
describe the enemy in 28 Days Later as zombie-like as they are
not so much undead as infected with a terrible virus.

What is a swarm?

A swarm is a whole that is more than the sum of its parts, bur it is also a heterogeneous whole. In the swarm, "the parts are not subservient to the whole – both exist simultaneously and because of each other."[32]

How does a swarm operate? Which is to say, how do individuals with minimal means to act on their own gain the ability to perform powerful actions when teamed with a large group of similar agents? This is a hot research topic, notable for its back and forth between the purely theoretical and mathematic, and observations of the natural world. We'll start there – in nature.

Large numbers of birds can often be seen flying in swirling clusters – coordinated, mesmerizing clouds. These swarming flocks are complex to model, and the processes going on are not always intuitive. [fig. 22] In the swarms, non-verbal communication flows through the cluster in waves. It would be easy to assume this process would increase in potency in a linear manner as the number of birds grew but it is far from that simple. Recent research has shown two different types of information waves are at play. Previously it was thought the swarming behavior emerged as the birds worked to maintain a single property: the distance between itself and the birds around it; new analysis suggests "information might travel more effectively as waves in flight orientation or "spin.""[33] Not only do the birds maintain both the concentration of birds and the direction they are flying, these two states of things operate on different scales, "only spin waves propagate in small flocks, whereas density waves dominate for large

---

[32] Berardi, Franco. *The Soul at Work: From Alienation to Autonomy.* Los Angeles, CA: Semiotext(e), 2009. 195.
[33] Schirber, Michael. "Synopsis: Silent Flocks." Physics. May 27, 2015.

flocks."[34] These different methods of swarm behavior emergence also result in an inverse goldilocks syndrome: at certain 'just right' numbers of birds, swarming behaviors fails, with neither concentration nor direction information successfully shared. This type of unexpected result is key to the study of complex systems that hide their tumultuous mechanisms beyond the easily visible and deducible. Natural phenomena provide a real-world laboratory in which to test a combination of theories and subsequent computer-driven models, with potential application in designing complex systems.

In a recent interview biologist Deborah Gordon describes the interdisciplinary relationship between her research on ant colonies and computer science. For instance, algorithms describing how ant forage, ensuring a sort of cost-benefits balance as to when its best for the colony to expend energy/water to acquire food have a lot in common with the Transmission Control Protocol regulating internet data traffic. Gordon and others' research into the methodology of natural swarms is immediately relevant in efforts to create systems of robots working as a team

In robotics, there's now a lot of interest in using the cheapest robots that require as little information as possible and that work together. A system like this is more robust to failure. Rather than sending one really complex robot to explore Mars or to search a burning building, it makes sense to send a group of cheap robots that will still work as a group even if one malfunctions. There are probably many new algorithms that ants have evolved to solve problems like this that we haven't thought of. What we need to do is go take a look.[35]

---

[34] Ibid.
[35] Singer, Emily. "Decoding the Remarkable Algorithms of Ants." Quanta Magazine. June 15, 2015.

An intermediary step between specialized robotics and a more general swarm is being developed by Google, albeit much larger and diffuse than what we imagine when we think of the term 'swarm'. Having acquired Titan Aerospace, their 'Project Titan' is a network of large solar-powered drones designed to provide internet service in rural and other not currently serviced areas. With a wingspan bigger than a Boeing 737 but weighing less than a passenger car, the idea is to keep the machines aloft for long periods of time. The project shares a goal with their balloon-based 'Project Loon'; each requires a 'simple' basic unit to be deployed en masse, working as a single net to maintain consistent coverage over a given area. In keeping with the company's general practice, the idea has dual philanthropic and commercial potentials, providing much needed access to an underserved population and unfettered contact to a new customer base wholly dependent on Google's products. But additionally, if such a service overcomes the many technical hurdles before it (maintaining a fleet of solar-powered constantly airborne vehicles is no simple feat), its hard to imagine the machines not including cameras, paving the way a new type of aerial surveillance at a much lower price point than satellites. Furthermore, it is easy to imagine the battlefield version invisibly elevated, perpetually aloft surveillance providing both intelligence and logistics/communication services to ground-based or drone forces.

Another practical application for these algorithms in concert with a multitude of robots is starting to emerge at a smaller, more familiar scale, filling a void left by the exit of another kind of swarm. Colony Collapse Disorder (CCD) was a term coined in 2006 to describe the phenomena whereby 20-40 percent of bee colonies had begun to die off each winter. Honey is only a small aspect of bees' importance. Pollination by bees is key to the "production of 39 of the world's 57 most important monoculture crops"; though those crops aren't necessarily staple foods the threat of bee extinction is a direct threat to human food cultivation.[36] This crisis is being addressed from multiple angles with scientists studying the cause of CCD (not singular, a mixture of pesticides, mites, fungus, malnutrition, immune deficiencies etc.) and beekeepers modifying their practices. A perhaps overly optimistic article's title suggests we "Call off the bee-pocalypse: U.S. honeybee colonies hit a 20-year high" as the latter, adjustments to beekeeping practices, seems to have stemmed the tide of losses, at least in the US.[37]

Nevertheless work is being done to deal with a temporary or permanent reduction in the number of bees: an effort to replace/augment their services with those of robots. Researchers at Harvard are developing tiny flying robots half the size of a paper clip that could act as pollinators. In 2013 the team demonstrated the first controlled flight of one of these tiny fly-like robots; work continues as the robot becomes lighter, stronger, faster, more durable, and eventually, capable of being released en masse into a flowering field. [fig. 23] In an article by the researchers for Scientific American, the researchers note

> RoboBees will work best when employed as swarms of thousands of individuals, coordinating their actions without relying on a single leader.[38]

---

[36] Williams, Geoffrey R., David R. Tarpy, Dennis Vanengelsdorp, Marie-Pierre Chauzat, Diana L. Cox-Foster, Keith S. Delaplane, Peter Neumann, Jeffery S. Pettis, Richard E. L. Rogers, and Dave Shutler. "Colony Collapse Disorder in Context." *BioEssays*, 2010, 845-46.

[37] Ingraham, Christopher. "Call off the Bee-pocalypse: U.S. Honeybee Colonies Hit a 20-year High." Washington Post. July 23, 2015.

[38] Wood, Robert, Radhika Nagpal, and Gu-Yeon Wei. "The Robobee Project Is Building Flying Robots the Size of Insects." Scientific American. March 22, 2013.

These robots would have no need for the flower's nectar as bees do (pollination being the flower's reward for supplying this food) much in the same way robot factory workers don't require any number of benefits unrelated to the task that their human counter parts do – we'll discuss this type of robotic labor solution in greater detail later. Key here is the image of this type semi-autonomous swarm – imagine the addition of the ability of the individuals to cooperate, to adjust their distribution to suit a particular crop, to later form meta-structures of multiple bots to, say, fly in and harvest the bot-pollinated crops...[39]

Gordon is currently with another scientist to understand the routes ants take in the complicated and ever-changing environment of a jungle. The study is leading to a non-intuitive realization: the ants do not aim to discover the shortest and most efficient path between their nest and a source of food through some sort of collective learning. Instead, they stumble upon *a* path, a potentially non-ideal but workable way to get where they need to go and they use it unflinchingly until a new path is required. This balance between optimization and minimal functionality runs counter to the instinct to find a best answer from the outset, a specialized solution to an isolated problem. Much effort in robotics up to this point has focused on the creation of extremely specialized machines particularly suited to a single or small array of tasks. In *Terminator* as in the military, a general-purpose soldier robot is the basic unit of the army, mirroring the way that in the army everyone gets 'basic training' whether one is preparing to drive a tank or drone, be a nurse, or work in a command role. But in Skynet's army, despite the potential to augmented a given robot with slightly different weaponry, the central skeletal humanoid form is still

---

[39] Incidentally I'll discuss a type of flying robot swarm that occurs in another story by Lem, *The Invincible*, a bit later.

rather specific, which is to say, not as generalizable as possible (versus the liquid metal creatures to follow).

Another way of thinking about solving problems in robotics is to think in terms of 'self-reconfiguring modular robots', generally speaking a (potentially) infinitely scalable group of simple robots (at least as far as their ability as individuals) that can interact with their kin to form ever more complex forms and complete more convoluted tasks − "from angular chaos, to robot-enabled order"[40]

Indeed, in Radhika Nagpal's Harvard laboratory (the same place as where the robot bees are being developed, a super simple robot called the Kilobot has been developed. Individually they aren't much, a small hockey puck shaped body with three stiff metal 'legs' which allow the bot to move by simply vibrating much as a cell phone will crawl a bit on a table when made to vibrate. But together, they can work to form shapes, perform tasks.

> The beauty of biological systems is that they are elegantly simple—and yet, in large numbers, accomplish the seemingly impossible," says Nagpal. "At some level you no longer even see the individuals; you just see the collective as an entity to itself.[41]

It is easy to imagine a slow progression towards "swarms of self-assembling microbots capable of reconfiguring themselves into different forms, shapes and sizes, and changing their function accordingly"[42] This echoes one of many predictions

---

[40] Lomas, Natasha. "MIT Scientists Create Modular Robot Blocks That Can Self-Assemble & Reconfigure." TechCrunch. October 4, 2013.

[41] Perry, Caroline. "A Self-organizing Thousand-robot Swarm." A Self-organizing Thousand-robot Swarm. August 14, 2014.

[42] Ibid. See also Sklar, Julia. "Diverse Robots Talk and Team Up to Complete Tasks for Humans | MIT Technology Review." MIT

discussed in reference to the 'singularity', a term popularized by science fiction writer Vernor Vinge among others describing a time in the near future "where our old models must be discarded and a new reality rules" due to rapid advances in technology.[43] In an overview of some of the paradigm-shifting technologies to come, author Annalee Newitz notes "Another singularity technology is the self-replicating molecular machine, also called autonomous nanobots, "gray goo," and a host of other things", an idea which sounds eerily close to our liquid metal friend.[44]

In a glossary of sorts of metaphors for our relationship to strong AI including such terms as *Scaffolding*, *Search Party*, and *Star System*, author Nora N. Khan talks about *Swarm*, specifically swarm intelligence, one of many distinct types of artificial super intelligence that Nick Bostrom outlines.

> A grouping of many millions of minds, deeply integrated into a singular intellect, [...] organized by elegant rules, with each individual mental event an expression of the mind's overall mission[45]

This metaphor could be used to describe humankind, the internet, and the structure of neural-net computing, albeit each with their own distict form of 'elegance'. It is along these lines that the T-1000 finds itself more sentient than its

---

Technology Review. August 4, 2015. Parallel to these developments are efforts to maximize the effectiveness of different robots with particular skills working together as a unit through robot-to-robot communication.

[43] For a very easy introduction to the idea of the Singularity see: Newitz, Annalee. "What Is The Singularity And Will You Live To See It?" Io9. May 10, 2010.

[44] ibid.

[45] Khan, Nora N. "Towards a Poetics of Artificial Superintelligence." After Us. September 25, 2015.

constituent parts, all abstract flecks of gray goo, could ever be on their own.

To understand the swarm mind is to understand all the component sub-wills, working in unison to create a burgeoning intelligence, something greater than the sum of its parts. As a swarm interacts and grows, despite no noticeable change in the intellect or sense of self of the individual parts, something close to consciousness approaches, emerges. Individual modules of the collective architecture line up with each function: learning, language and decision-making.[46]

Bostrum suggests that the swarm has the latent potential to, having reached a critical mass, form a kind of consciousness. The previously cited theorist Kevin Kelly describes this process:

> As very large webs penetrate the made world, we see the first glimpses of what emerges from the net-machines that become alive, smart, and evolve – a neo-biological civilization. There is a sense in which a global mind also emerges in a network culture. The global mind is the union of computer and nature – of telephones and human brains and more. It is a very large complexity of indeterminate shape governed by invisible hand of its own.[47]

Hives and swarms... the degree to which a given metaphor or does or doesn't fit its target becomes a project/process rather than a key. To Khan,

---

[46] Ibid.
[47] Kelly, Kevin. Out of Control: The New Biology of Machines, Social Systems, and the Economic World. Reading, Mass.: Addison-Wesley, 1995. Read in
Berardi, Franco. The Soul at Work: From Alienation to Autonomy.

better suited poetics could be a form of existential risk mitigation. Using metaphorical language that actually fits the risks that face us means we will be cognitively better equipped to face those risks.[48]

When risks are legion, to focus our energy on any single actor (or worse still, a single descriptor) is like attacking a horde of locusts with a pistol.

---

[48] Ibid.

> We are not there yet. Indeed, TrueNorth is a *direction* and
> not a *destination*! The end goal is building *intelligent business*
> *machines* that enable a cognitive planet, while *transforming*
> *industries*. Exciting!
> *-IBM Fellow Dharmendra S. Modha, speaking about TrueNorth,*
> *a new form of brain-inspired computer chip. Emphasis is the*
> *author's.*

> Anything I've learned about giving computers autonomy
> is that they crash, so I'm not so much worried about
> them taking over as I am about the present in which they
> fail dramatically.
> *-MIT Media Lab researcher Kevin Slavin, at the 'Wired Money'*
> *conference in 2013*

I've spoke at length earlier about some of the potential bad
turns artificial intelligence can – and frequently does – take in
fiction based on errors in programming. In these stories the
initial directives given non-biological intelligences are
misinterpreted or taken to an extreme which causes a
powerful AI system/entity to act in a way detrimental to its
creator and/or humanity.

In the way human ailments vary dramatically in scale from a
scratch to a broken leg to parasitic infestation to viral
infection to cancer, software-based entities are capable of
failing on any number of registers. A simple, unchanging rule
structure is relatively immune but can itself suffer from the
introduction of noise into the system from interference in its
internal signaling structure. There can also be failures of the
physical stuff of which even the most advanced system must
still be made, its storage medium, power grid et al. It is easy
to forget as we move from the moving parts-based hard drive
to solid-state storage, and we slowly move from local storage
into the cloud, that there remains always a core amount of

matter somewhere, subject to entropy. A good system works in various manners of redundancy and checks and balances to keep a small error from spiraling outward to effect the whole structure but never the less the physical objects at the base of it all have their own associated risks from heat-based failure to the ever-present peril of a spilt glass of water.

Generally the every day code elements of software have in place protections which give us a reasonable level of security that the whole will not be compromised. Our cells have analogous methods for maintaining their own code, DNA, the contents of which are also subject to error. But despite a constant effort to maintain order, errors do crop up and occasionally rise to the level of existential threat. Just like the knowledge of the concept of 'cancer' fails to describe the wide scope of typologies and causes nor does it adequately lead to ways to diagnose and cure, so too will increasingly large digital systems need continued research to identify, 'cure' and prevent small-scale errors from proving malignant.

Upon reaching the level of complicated, learning machines in which an initial set of rules is self-augmented through experience, these rule structures themselves become prospective sites for errors, potentially in completely unforeseeable ways. Omohundro in *The Basic AI Drives* gives a great example:

> Eurisko was an AI system developed in 1976 that could learn from its own actions. It had a mechanism for evaluating rules by measuring how often they contributed to positive outcomes. Unfortunately this system was subject to corruption. A rule arose whose only action was to search the system for highly rated rules and to put itself on the list of rules which had proposed them. This "parasite" rule achieved a very

high rating because it appeared to be partly responsible for anything good that happened in the system.[49]

A virtual Darwinism plays out in what appears to be a stable, self-contained digital system, a kind of nature (the initial state of the system) versus nurture (the addition of experience, data, mutation etc.) The challenge is maximizing the usefulness of these additions to a system over time while minimizing their potentially injurious effects. Humans are no different; we perceive a neat and tidy border between ourselves and other organisms when in fact we play host to a massive number of hangers on. Indeed bacteria and other non-human cells outnumber our own ten to one.[50] The majority of these beings are not only not harmful, they provide essential assistance to their human host, regulating other harmful bacteria and/or providing needed services such as helping in digestion etc. In a recent journal article by Peter Kramer and Paola Bressan in *Perspectives on Psychological Science* humans are described as

> not unitary individuals in control of ourselves but rather 'holobionts' or superorganisms—meant here as collections of human and nonhuman elements that are to varying degrees integrated and, in an incessant struggle, jointly define who we are.[51] [fig. 24]

As currently constructed, adapted to a certain level of dependency upon other species, a human would die fairly immediately if forced to operate alone, without these co-

---

[49] Omohundro, Stephen M. "The Basic AI Drives." Self-Aware Systems. 2001.

[50] For more on our strange community of creatures that make up a person: "NIH Human Microbiome Project Defines Normal Bacterial Makeup of the Body." U.S National Library of Medicine. June 13, 2013.

[51] Found in an article: Justin. "We Are Not Human Individuals." Daily Nous. July 24, 2015. Some infographics were included; they'll be added to the plates of this book.

passengers.[52] But these dependencies are not without risk. In the above example of Eurisko, there is nothing harmful about this particular 'corruption' – the rule had no negative consequences besides being purely extraneous. But it manages to persist using a mechanism via which a malignant rule would endear itself to the whole. A complex system through its very complexity becomes porous and subject to both attacks from outside and within.

From outside: bacteria both harmful and beneficial (and neutral) use the same methodology to gain access and find sanctuary in our body, and viruses use the cell's own method of reproduction to their own ends. To be clear, from an evolutionary standpoint an organism successfully harming or killing its hosts in any large numbers will be 'selected against' i.e. will not last long, running out of hosts or running afoul of their hosts ability to regulate via immune system or medical intervention such intrusions. Likewise learning systems by their very nature are constantly interacting with data beyond their initial programming, within which there will be a mix of the useless, useful, and malicious (intentional or otherwise). This mirrors human-to-human interactions: mostly innocuous, occasionally beneficial, and occasionally toxic.

From inside: systems trained to learn have to 'experiment' with different combinations of data, mining for generalization and trend in order for small changes to echo through the whole

---

[52] Donna Haraway puts this well:
Some of these personal microscopic biota are dangerous to the me who is writing this sentence; they are held in check for now by the measures of the coordinated symphony of all the others, human cells and not, that make the conscious me possible. I love that when "I" die, all these benign and dangerous symbionts will take over and use whatever is left of "my" body, if only for a while, since "we" are necessary to one another in real time." Haraway, Donna Jeanne. When Species Meet. Minneapolis: University of Minnesota Press, 2008.

system. Imagine an AI mind as a kind of algorithmic biome whose stasis will inevitably find itself interrupted when one of its constituent inhabitant species falls victim to disease, allowing populations of others creatures normally kept in check to explode.

These internally created conflicts echoes the cancer metaphor used earlier – an unexpected (identity)crisis set in motion by a small disturbance such as a miss-copy etc. There is a chance that this type of small error might be even more potent and potentially devastating in large, hyper complicated systems in which the normal, predictable growth of connections between elements experiences 'explosive percolation'. Similar to metastasic disease in which cancer spreads from one organ or region to another seemingly disconnected one, explosive percolation is an emergent phenomenon of über connectivity, a type of special case in which a phenomena might find itself spreading rapidly due to sudden, massive connectivity; it "might emerge with a bang, not a whimper."[53] This can be caused by a number of factors including a slight tweak to global rules defining internal structures, or when

> human operators or regulators intervene with a network's functions or structure in an attempt to delay an undesirable outcome, such as a leak in a dam or a crash in a financial market. Such delayed interventions can sometimes be successful, yet at other times lead to unanticipated and disastrous failures.[54]

That a series of well meaning small-scale interventions can have dramatic consequences — for good or ill – suggests that

---

[53] Ouellette, Jennifer. "The New Laws of Explosive Networks." Quanta, July 14, 2015.
[54] Drawn from the conclusion; the article is not for the faint of heart aka is far too technical for me: D'Souza, Raissa M., and Jan Nagler. "Anomalous Critical and Supercritical Phenomena in Explosive Percolation." Nat Phys Nature Physics, 2015, 531-38.

there is the potential to maliciously provoke such unpredictability. Or as detrimental, efforts to avoid disaster might themselves result in something worse. Physicist Raissa D'Souza who co-authored a recent paper on the subject describes the trajectory of the research: "We'd like to be able to intervene in the system easily to enhance or delay its connectivity"[55] Explosive percolation is a first step in thinking about control, according to D'Souza, because it provides a means of manipulating the onset of long-range connectivity via small-scale interactions. But these acts themselves have the potential to similarly put a 'finger on the scale' so to speak – are themselves actions which might have their own unexpected results.

And overly limited connectivity may fatally restrict the usefulness of a system. The Russian cybernetics project of the early '60s as led by the Cybernetics Council of the Soviet Academy of Sciences failed exactly out of an effort to avoid this kind of overreaching and uncontrollable interconnectivity. Slava Gerovitch argues that the project of "applying computers and cybernetic models in a wide range of fields, from biology and medicine to production control, transportation, and economics" fell victim to not a lack of technical ingenuity on the part of Russian scientists, but to mismanagement by the various heads of governmental agencies.

> Cyberneticians hoped to establish a new central agency to oversee information management in all other government bodies, but individual ministries succeeded in appropriating the role of primary users of management information systems. [...] fracturing the network into unconnected islands.[56]

---

[55] Ouellette, Jennifer. "The New Laws of Explosive Networks."
[56] Gerovitch, "InterNyet: Why the Soviet Union did not build a nationwide computer network"

Over regulation by multiple agents resulted in a 'too many cooks in the kitchen' situation, a system built to achieve broad goals unable to achieve them due to its individual elements being too empowered to control connectivity and prioritization. Without a balanced approach to connectivity, building the infrastructure to connect complex systems is a wasted effort, unable to achieve any of the improvements in productivity intended by such a project.

Balancing external control schemes with the necessary autonomy with which networks are their most production is no easy task. A nightmare of connectivity of sorts (and a bit of a detour for this chapter, my apologies), *The Lawnmower Man* (Leonard, 1992) revolves around a researcher Dr. Lawrence Angelo applying a regimen of virtual-reality-based learning and brain stimulating drugs to a less-than genius-level local gardener named Jobe.[57] [fig. 25] The basic plot is very similar to the science fiction short story and novel by Daniel Keys *Flowers for Algernon* from the late fifties involving an experimented upon mouse named Algernon, and man with a very low IQ, Charlie Gordon, each of whose intelligence is significantly altered. Whereas Algernon and Gordon's new intellect is tragically fleeting, Gordon losing the privileges and relationships gained while super smart, *The Lawnmower Man* forgoes eventual decline (at least intelligence-wise). Instead Jobe becomes increasingly erratic, binges on both VR and the brain drugs, develops supernatural powers, and begins to mete out revenge to those he feels crossed him when he was

---

[57] It has to be mentioned that the title 'The Lawnmower Man' and the initial marketing of the film as *Stephen King's The Lawnmower Man* stems from an unrelated short story by King. The film rather roughly combined the slightest whiff of the original story with a screenplay by director Brett Leonard and producer Gimel Everett called 'Cyber God', hoping to capitalize on the popularity of King. After much litigation the reference to King was removed from subsequent marketing.

mentally inferior. It is worthwhile to draw a brief parallel between the earlier-mentioned b-horror film *The Brain That Wouldn't Die* (Green, 1962). In it, a mad scientist's fiancé is decapitated in an accident and her head is kept alive on a pan via special liquids of some sort being piped though various laboratory apparatus. Not only is her dismembered head somehow fully aware and able to speak, as she sits there wishing she were dead she begins to communicate psychically with one of the doctor's previous monstrosities locked away. In each of these films the approach to bodiless-ness is accompanied by psychic powers. It is as if there is a minimal human physical agency which, when repressed, can find its outlet beyond what is allowed with the normal bounds of logic and science.[58]

Dr. Angelo tries to stop Jobe; Jobe increasingly desires to become digital, to upload himself completely into the virtual world. Dr. Angelo tries to walk Jobe back from the precipice overlooking unredeemable madness, but Jobe is not to be dissuaded (from the script to the film):

> Jobe: I'm going back to VSl to complete the final stage of my evolution. I'm going to project myself into the mainframe computer. I'll become pure energy. Once I've entered in the neural net... my birth cry will be the sound... of every phone on this planet ringing in unison.

Much effort is made to stop Jobe both physically and virtually, culminating in Dr. Angelo destroying the laboratory

---

[58] To quote the mad scientist from The Brain That Would Not Die film:
The paths of experimentation twist and turn through mountains of miscalculations and often lose themselves in error and darkness!" And that darkness is that which is beyond the laws of the physical world; this fall-back is something worth writing about another time.

housing Jobe, the VR system, and computers used in the experiment in a dramatic explosion. Everything seems to have ended happily ever after, with the scientist recording an audio journal entry:

> Dr. Angelo: Last journal entry for a while. I won't let Jobe's death be for nothing. What happened to him is my responsibility. For some reason, I've been given a second chance so I'm taking my work underground. I can't let it fall into the wrong hands again. If we can somehow embrace our wisdom instead of ignorance this technology will free the mind of man... *(pauses lighting a cigarette to stair at the match's flame)*
> ...not enslave it *(blows out match without lighting cigarette)*

But this measured optimism is short lived. On the way out to presumably go on the run and 'underground', his phone rings; you can hear another phone in the distance, then another. The camera shows a series of skylines, each accompanied by the sound of countless phones ringing, and the film ends.

In *The Lawnmower Man* the supernatural element plays second fiddle to Jobe's resources in the digital domain and his eventual departure as 'pure energy', which is to say 'pure data'. The nightmare is not the reasonably horrifying idea of a homicidal maniac with telekinesis and super intelligence running free through the city. Rather, it is that same super intelligence loose in the digital realm, capable of being anywhere, or everywhere, which renders the final scene as effective.

It works at least in part because the scenario seems relatively realistic, at least at face value. We lack specialized knowledge in regards to technology – are the world's phone systems so very interconnected, addressable in unison? Though likely not the case it at least seems plausible and recalls the aforementioned potential for malicious use of über connectivity. And works as a thought experiment to emphasize the relevance of checks being placed allowing the free flow of information to be regulated – a toolset as dangerous in the wrong hand as any weapon but never the less worth pursuing.

Virtual reality is also a 'realistic' thing, not so much an invention as a series of interconnected technologies which when combined present a user a convincingly immersive virtual experience. When this idea was popularized in the nineties it was so far from perfect that it felt like an unredeemable failure. In subsequent years it disappeared from public view (and presumably back to the drawing board), seeming almost quaint when brought up in pop culture, only to see it recently reemerge with multiple consumer-level virtual reality products suddenly on the market. As should have been expected by virtual reality's naysayers, technology has advanced significantly, and will continue to do so as the commercial potential of VR finally begins to be realized. There are few fields of human endeavor not currently exploring some manner of VR future, from therapy, medical surgery via tele-presence, and both obvious (video games) and less obvious (television news) media. Professor Bob Stone, chair of interactive multimedia systems at the University of Birmingham who has been working with VR since its initial boom and bust notes:

Immersivity is the main thing. This has many years to run, and we have to get to where we are totally convinced – but we're looking at something which has transformational capability for society.[59]

---

[59] Arthur, Charles. "The Return of Virtual Reality: 'this Is as Big an Opportunity as the Internet'" The Guardian. May 28, 2015.

Jobe's eventual success uploading himself into a purely digital entity, maybe the most outlandish element, echoes one of many predictions discussed in reference to the previously discussed singularity. For Jobe, assuming there was enough memory and bandwidth to somehow upload his mind (certainly at the time there was neither), it is hard to imagine he having the equipment on hand to perform the deed. It is proposed by futurist Raymond Kurtzweil and others that in the next fifty years or so due the law of accelerating returns, it will be technically possible to achieve this kind of virtual immortality. We will be able to map the entirety of our brain's chemical and electrical organizations, allowing our minds to be digitized Jobe-like and to then be duplicated and uploaded, in some manner thriving in the digital realm and/or be transferred into new physical bodies.

Science fiction author Ursula le Guin, poetically discussing the internet on her blog, offers

> To be free of the body, tied to no place in time and no time in place, yet having effortless, limitless access to everyone one knows, to all knowledge, and to immediate or securely promised satisfaction of desires would appear to be the condition of a blessed immortality.
> What could possibly be wrong with it?[60]

It is hard not to read the 'What could possibly go wrong?' as either tongue-in-cheek or naïve; perhaps it is both. The believers in the singularity propose to actualize what is currently only imagined in metaphor. The likelihood of their predictions is a subject of much debate, with opponents for example questioning the reliance on the application of exponential growth to systems that may favor logistic growth, tapering off after a while. And implicit in the singularity's realization is the glossing over of the difference between

---

[60] Ursula le Guin's blog June 2, 2014

"biological data collection with biological insight"[61]. It is not enough for technology to allow for the creation of larger and larger datasets, the algorithms which interpret and make use of these datasets must also advance in kind.

It is no surprise that proponents overlook possible speed bumps on the way to the singularity. The majority of apocalyptic fantasies place the end of days just beyond their lifespan/horizon, while most techno-utopic visions place the time of achieved techno-immortality just within their grasp, corresponding with their own life expectancy.[62] And the way progress thus far is used to extrapolate future success more often than not makes the predictor seem silly in hindsight. A New York Times article in 1958 began with the following optimistic predictions, most of which have yet to coalesce into a single object/entity:

> The Navy last week demonstrated the embryo of an electronic computer named the Perceptron which, when completed in about a year, is expected to be the first non-living mechanism able to perceive, recognize and identify its surroundings without human training or control [...] the embryo of an electronic computer that [the Navy] expects will be able to walk, talk, see, write, reproduce itself and be conscious of its existence.[63]

These flaws aside, there is insight to be gained by looking at visions of the future, both fictional and non-fictional, though the later may be inextricably tied to individual ambitions,

---

[61] Amyx, Scott. "Wearing Your Intelligence: How to Apply Artificial Intelligence in Wearables and IoT." Wired.com. December 2014.
[62] Kurtzweil is 67 and expects the singularity in 2045, when he'll be 97; he takes an intense regimen of vitamins and other heath-related steps to stay alive long enough for his version of immortality to arrive.
[63] Found in the wiki for Perceptron.

fears and desires posing as impartial predictions based on facts.

As Jobe is becoming unhinged, he begins to see himself more and more as a God, something the scientist chides him about what he considers delusions: "Listen to what you're saying. The first sign of psychosis is a Christ complex. Cyberchrist." But in the final scenario, through not entirely outlandish means, Jobe does what he sets out to do and becomes a bit of a god, omni*present* at least if not omni*potent* (phone-ringing skills not withstanding), and this achievement via 'practical' means is frightening.

It is similar to the Y2K or millennium bug, a kind of digital 'act of God' that never truly came to be. Deep inside all computers, from ones used for homework to ones running power plants and weapons systems there was a core inability to deal with the turn of the date number from 1999 to 2000 which had to be patched. This inspired a number of doomsday scenarios as it seemed impossible that all relevant machines would be repaired in time. Connectivity was again here the nightmare – should one link in the chain fail it was thought its errors would cascade across systems leading to power outages, stock market failures and other unforeseeable catastrophes. This never came to be; over 300 billion dollars were spent (134 in the US alone) prior to the roll over and the day came and went with no noticeable disturbances. Furthermore, in sectors such as education and small businesses where preparedness was lacking, there were no major problems not solvable on an as-needed basis.

This 'failure to fail' notwithstanding, the mere conjuring to mind of computers as a homogenous group capable of failing en masse, or in Jobe's case, being controlled by a single entity, foregrounds our growing dependence on computing, and the threat of a cascading error across the globe. Whereas we can imagine – usually through gross oversimplification but imagine nonetheless – the inner workings of other necessities like electricity and water, piped into our homes and ready to use, there is an added level of abstraction to digital systems which seems to increase as they become more ubiquitous and entrenched, especially as we begin to take steps to enmesh the physical and digital worlds – an internet of things.

Bostrom in a paper about global existential risk describes "a black ball: an easy-to-make intervention that causes extremely widespread harm and against which effective defense is infeasible"[64] In addition to real 'black balls', imaginary ones have the power to both cause fear and to call for solutions to be devised lest they turn actual. Additionally, they draw to the surface through the exaggeration to the point of cataclysm under discussed discomforts with the state of things, where they seem to be going, and suggesting models for preemption – or inevitability.

---

[64] Bostrom, Nick. "Existential Risk Prevention as Global Priority." *Global Policy* 4, no. A (2013): 15-31. Bostrom somehow misses the irony behind one of his 'policy implications':

> Perhaps the most cost-effective way to reduce existential risks today is to fund analysis of a wide range of existential risks and potential mitigation strategies, with a long-term perspective.

Which is essentially to say, the best hope for the long term survival of humanity is more funding for his own research, which, though it may be true, feels a little self-aggrandizing. Nevertheless, his text is a good read, connecting various speculations about the future of our species within a single framework.

John Perry Barlow: It's not that there's anything particularly healthy about cyberspace in itself, but the way in which cyberspace breaks down barriers. Cyberspace makes person-to-person interaction much more likely in an already fragmented society. The thing that people need desperately is random encounter. That's what community has.

bell hooks: Seeing your computer, it feels like this lively possibility where anything may flash itself on that screen. *-From a wide-ranging conversation between John Perry Barlow and bell hooks published in Buddhist magazine Shambhala Sun.* [fig. 26]

The internet, despite being initially a military project, came into its adolescence with endless promise, a virtual world that would make "person-to-person interaction much more likely", with the potential for "anything to flash itself onto that screen", versus feeling alienating in the former and potentially embarrassing/offensive in the latter cases. The internet initially

favored military values, such as survivability, flexibility, and high performance, over commercial goals, such as low cost, simplicity, or consumer appeal

but its is these core values which made it a robust framework capable of endure the wild success and growth which would eventual make it such a fundamental part of the contemporary world.[65]

---

[65] Abbate, Janet. Inventing the Internet. Cambridge, Mass: MIT Press, 1999. 5.

The Telecom Reform Act of 1996 sought to control and censor content on the then nascent internet, much to the chagrin of the its small but vocal user base at the time, a community of early adapters – the 'citizens of cyberspace'. Writer John Perry Barlow penned a text in response he hoped (in the email introduction to a broader rebuttal to the text) would "echo across Cyberspace, changing and growing and self-replicating, until it becomes a great shout equal to the idiocy they have just inflicted upon us."[66]

> Governments of the Industrial World, you weary giants of flesh and steel, I come from Cyberspace, the new home of Mind. […] You do not know our culture, our ethics, or the unwritten codes that already provide our society more order than could be obtained by any of your impositions. […] Cyberspace consists of transactions, relationships, and thought itself, arrayed like a standing wave in the web of our communications. Ours is a world that is both everywhere and nowhere, but it is not where bodies live.[67]

The text is a messy affair – idealistic, bombastic, with his (in his own words) "characteristic grandiosity"; it describes a world "soon be blanketed in bit-bearing media." Cyberspace was seen as the very model of a libertarian, self-governing

---

[66] Barlow, John Perry. "A Declaration of the Independence of Cyberspace." A Declaration of the Independence of Cyberspace. February 8, 1996.
I used the descriptor 'writer' as an expedient; in his wiki he is described as an

> American poet and essayist, a retired Wyoming cattle rancher, and a cyberlibertarian political activist who has been associated with both the Democratic and Republican parties. He is also a former lyricist for the Grateful Dead and a founding member of the Electronic Frontier Foundation and Freedom of the Press Foundation.

[67] Ibid.

utopia, "without privilege or prejudice accorded by race, economic power, military force, or station of birth."

Fellow theorist and founding executive editor of Wired Kevin Kelly wrote a long, wide ranging piece for New Age Journal in 1984 called *The Birth of a Network Nation [plate x]*. In it he describes his experiences on early message boards and communities called Electronic Information Exchange Systems, or EIES, including groups dedicated to cyber activism, tele-education, human-based search engine-like librarians, virtual soap operas, cyber religion, and early 'social networking' ("The thing I grow to love about EIES is that it is a place to meet, perhaps like the cafes of old Paris").[68] He spoke of the phenomena of LOL'ing (not using the term though it was invented around that time, supposedly on a Canadian bulletin board system by a certain Wayne Pearson though records are scarce)

> The laughter is what surprises me most. I often find myself in my room at night, reclining back in my chair, reading communiqués as they scroll by on the green monitor, and laughing deep belly laughs at the wry humor. I rarely do that reading. I'm more surprised to hear other teleconferees confess the same. I e-mail (electronic-mail) a question to Turoff. What does this uncommon laughter mean? I figure he should know since he spends eight to ten hours a day hooked up to his own creation. "It's about letting down face," he says. 'We did some experiments with one-way mirrors and found that when people are on-line, they let down their face, that is, they drop their guard more than usual. It's a telephenomenon.[69]

---

[68] Kelly, Kevin. "The Birth of a Network Nation." *New Age Journal*, October 1, 1984, 42.
[69] Ibid.

Kelly imagined the rise of a neo-biological civilization online, managed by bottom-up-structured, networked systems outside of traditional hierarchic control schemes. At the time the grass roots, bottom up nature of the web had an exciting vitality, specifically the ability for "special-interest and advocacy groups, the socially disadvantaged, rural citizens, and kids" to set up local networks:

> The socially disadvantaged, a term encompassing almost anyone from the disabled to the elderly, can communicate and receive equal respect. Rural populaces become less remote. Kids thrive in the hidden electronic corners of the network nation. Networking diminishes image and lessens power and roles where image is important. Instead, it rewards ideas.[70]

A 1994 piece for Harper's pitted two opposing views against each other in a pair of opinion pieces called "The Electronic Hive: Two Views." Kelly writes in his piece, titled "Embrace It" about the icon of science of the twentieth century, the Atom, being rapidly replaced with the Net.

> Like the beehive, the Net is controlled by no one; no one is in charge. [...] A recurring vision swirls in the shared mind of the Net, a vision that nearly every member glimpses, if only momentarily: of wiring human and artificial minds into one planetary soul. This incipient techno-spiritualism is all the more remarkable because of how unexpected it has been. [...] In the process of connecting everything to everything, computers elevate the power of the small player. They make room for the different, and they reward small innovations. Instead of enforcing uniformity, they promote heterogeneity and autonomy. Instead of sucking the soul from human bodies, turning computer-users into an army of dull

[70] ibid

clones, networked computers -- by reflecting the networked nature of our own brains and bodies -- encourage the humanism of their users. Because they have taken on the flexibility, adaptability, and self-connecting governance of organic systems, we become more human, not less so, when we use them.[71]

Author Sven Birkerts wrote the opposing piece, titled simply "Refuse It". His text reads as very prescient, rife with predictions of many of our contemporary anxieties:

I am startled, though, by how little we are debating the deeper philosophical ramifications. [...] Why do we hear so few people asking whether we might not *ourselves* be changing, and whether the changes are necessarily for the good? [...] We have created invisible elsewheres that are as immediate as our actual surroundings. We have fractured the flow of time, layered it into competing simultaneities. We learn to do five things at once or pay the price. Immersed in an environment of invisible signals and operations, we find it as unthinkable to walk five miles to visit a friend as it was once unthinkable to speak across that distance through a wire.[72]

Kelly has since revised his idealistic view of cyberspace organization, suggesting that 'the bottom is not enough. You need a bit of top-down as well.'[73] Barlow's optimism has waned a bit as well in the decades since the nineties; when

---

[71] Birkerts, Sven, and Kevin Kelly. "The Electronic Hive: Two Views. (computer Networks and Other Digital Communication)." *Harper's Magazine*, May 1, 1994.
[72] Ibid.
[73] Kevin Kelly, "The Technium: The Bottom Is Not Enough", February 12, 2008, http://www.kk.org/thetechnium/archives/2008/02/the_bottom_is_n.php. This section is indebted to Annabel Frearson's article 'Infomanticism' in the magazine *QGJCPLB*, published by FormContent.

asked why he put it simply: "We all get older and smarter."[74] I think it is telling that even the term *cyberspace* itself feels antiquated, as behind the times as enunciating the *www* at the beginning of a url. What has happened to this notion of an endless, expansive space? We have settled quite firmly onto the internet, an *inter*connected *net*work, internal and between, a one-dimensional point-to-point and linear space versus a vast multi-dimensional plane. An 'answers.com' entry by Rankiri on the difference between cyberspace and internet notes:

> Practically speaking, it probably makes more sense to leave "cyberspace" to science fiction writers and use less pompous "Internet" and "World Wide Web" for descriptions of one's everyday online experiences.

Yahoo answers uses "desertcities" ends his entry – which fondly recalls such transitional phases from ARPANET such as USENET and BITNET – suggests

> I guess you could think of planet earth as the internet and then look upward at the unlimited stars, far reaching, out there, but quite real. That would be cyberspace to me.

Why have we left cyberspace for the science fiction writers, that great field of unlimited stars beyond?

Referring back to Kelly's 'Embrace It' text, he goes a bit further with his beehive metaphor

> The tiny bees in a hive are more or less unaware of their colony, but their collective hive mind transcends their small bee minds. As we wire ourselves up into a hivish

---

[74] Doherty, Brian. "John Perry Barlow 2.0." Reason.com. August 1, 2004.

network, many things will emerge that we, as mere neurons in the network, don't expect, don't understand, can't control, or don't even perceive. That's the price for any emergent hive mind.[75]

The same metaphor starts to show it dark side today in Yann Moulier Boutang's question "Are we all just Google's worker bees?"[76] That seemingly immaterial hiving without a single purpose is, to a well-placed entrepreneurship, a minable, free labor force. Google's algorithms are incomplete things, incapable of solving an individual search query without the massive, ever morphing database of our collective internet usage. People represent both the consumer and producers – we are the customers, Google's system is the middle management and we are in a strangely circular manner also the laborers. Think of the benefit from collecting not only our search history but the style with which we use the pages we get to – certain people lingering long on each page while others selectively spending their time on this of that page; some more or less likely to either go back to the initial search to try another link versus taking a first-choice several levels deeper. Qualifying the type of internet user by style in this way is only one of many potential kinds of information offered up by a user on a regular basis, with only this information's utilization to be decided. For a user that tends to be tied to the veracity of their search's initial results, potentials ads would be best spent spread across the top bunch of results; for those likely to click the top link and then journey out from there, it would be better to create ads tailored to appear interspersed out among the various likely branches of a search. The incidental production of our online activity, the massive database which only grows in complexity and nuance, is akin to bee's main service – "Beekeepers in the

[75] Ibid.

[76] Castiglione, Chris. "Yann Moulier Boutang Asks, "Are We All Just Google's Worker Bees?"." Institute of Network Cultures. November 13, 2009.

U.S. are no longer making their living by selling wax or honey. They are selling the bee's activity: they rent their service of pollination."[77] This sentiment is echoed in Robin Mackay & Armen Avanessian's introduction to *#ACCELERATE: The Accelerationist Reader* "Seen from the future, might the human prove nothing but a pollinator of a machine civilization to come?"[78] Our daily internet usage is the honey, the training we are giving countless learning algorithms the pollination.

In order for the 'citizens of cyberspace' to be kept productive and 'citizens of the state/capitalist system', the internet's boundless potential has to be constantly muted, simplified, centralized. It is grounded increasingly in the everyday, in the mundane, having become a utility. In his analysis of innovation, buzzword par excellence for what tech firms specialize in, ethnologist Marc Augé highlights the opposition between this term and something it is confused with – creativity. Innovation is at its core about the transformation of ideas into products. What dreaming and creativity were to the initial phases of cyberspace, products and innovation are to the internet, they are what keep it grounded so far from those 'unlimited stars.'

Augé reminds us that, counter-intuitively, so much of what counts as innovation springs from the minds of consumers, their ideas given freely to those with the means of production precisely so that they (users/innovators) can buy their own ideas back, standardized, sterilized, traceable. Innovation is another field in which consumption and production have been intertwined; that this process of 'distributed innovation' is often referred to as 'democratization' serves as an

---

[77] Ibid.
[78] Mackay, Robin, and Armen Avanessian. *#ACCELERATE: The Accelerationist Reader*. Falmouth: Urbanomic Media, 2014. 6.

indictment of the term rather than any kind of moral endorsement of the state of things.

bell hooks' "lively possibility where anything may flash itself on that screen" became the pop-up, something to be banned and or blocked. Or it is our social media stream – a place where surprises still strike one but only after they've been thoroughly vetted by the complex algorithms of learning software as not containing whatever taboos (violating social mores but also treading on copyright owner's feet) the site moderators are inclined to enforce.[79]

The internet feels infinite but in fact its edges exist; it doesn't take much effort to think of information not easily accessed (for instance a particular song will fall outside the purview of Spotify or YouTube). But its scale, however trimmed, censured, bandwidth-capped, region-locked etc. is beyond that of a single user, and perhaps were every human to turn their attention to reading and viewing all of the internet's content the task would still prove undoable. Boris Groys describes who the real viewer is online:

> But here the following question emerges: who is the spectator on the internet? The individual human being cannot be such a spectator. But the internet also does not need God as its spectator – the internet is big but finite. Actually, we know who the spectator is on the internet: it is the algorithm – like algorithms used by Google and the NSA.

---

[79] Surprises via social media occur less and less. I still laugh at a particularly nonsequitous post regularly, but as the software delimits more and more tightly whose content I see based on its opinion of what I want most these social sites become ever more aligned with my IRL community versus allowing for a broader conversation.

Of course, one can say that the algorithm cannot be seduced or frightened. However, this is not what is actually at stake here.[80]

This last sentence is shortsighted – algorithms are seduced and frightened all the time. When building a website one tries to construct the data both visible and hidden in code (metatags etc) so as to appeal to Google's search algorithms, to appear higher in a given search. And as softwares have begun to write basic articles based on the news, the drive for this content to gain human viewership and seem non-artificial is tempered by the whims of another type of interested party – these AI-authors must tailor their reportage and headlines to seduce other softwares, to appear high in search and to induce algorithm-driven reposting on semi-automated news sites.[81] And algorithmic fear is a constant as the web tries to

[80] Groys, Boris. "The Truth of Art." The Truth of Art. March 2016

[81] And AI is not focusing on just boilerplate reportage; it is trying its hand at creative writing as well. A text written by software managed to be accepted into a literary competition in Japan. Guided by researchers, the AI constructed "a novel entitled *Konpyuta ga shosetsu wo kaku hi*, or *The Day a Computer Writes a Novel*, about an AI that abandons its responsibilities to humanity after recognizes its own talent for writing." (Tarantola, Andrew. "AI-written Novel Passes First round of a Literary Competition." Engadget. March 24, 2016.) Charmingly meta, this story neither sprung into being fully formed (researchers guided the system's writing, giving it phrases and words to utilize) nor is the result of a software somehow 'deciding' to write a story on a whim. But what such an example suggests is those two caveats might be overcome – creative productions may emerge, if we allow in our programming of AI enough room for such improvisation.

And as intriguingly, one could image various productive processes of back and forth between machine and human. For instance, one could outline the barest bones of a novel, pass those elements through some manner of software which would then compose the full text. The human could go back in, soften the edges, add bits here and there and repeat the process. We do something like this

keep one step ahead of the endless bots and viruses trying at all times to hijack data; we are warned by our browser not to proceed opening this page, downloading this or that file etc.

The internet is battleground in constant flux. Having the best content is not enough; like any media or indeed any sport it is not enough to win the game on merit – the combatants also must make sure their play pleases the audience. In sports this often means perfect play is balanced against showmanship – a flashy high-risk dunk is valued more than an easy layup though they result in the same change to the score. On the internet the same battle for viewers plays out behind the scenes but is front and center in the digital realm as pages are read, ranked, shared, re-read, re-ranked, re-shared proportionally to their willingness to play this inside game.[82]

Indeed, as advertising dollars flow based upon views, how does one contend with code popular with code, the use value of human consumption increasingly secondary to analytic considerations?

It is at least partially true as Barlow says that "Cyberspace makes person-to-person interaction much more likely in an already fragmented society", for instance it is easier for me to

---

already – as I write this the computer gives me input such as the green underline hovering beneath the first sentence of this paragraph, or a red underline beneath a certain name. I can give the software feedback, letting it know *curatorial* is to be considered a word, *internet* is fine not capitalized etc. I ran this book through several online tools which summarize a given document. The results generally were complete gibberish, random bits of text smashed together. But here and there, relationships between sections I hadn't noticed emerged, links I reinforced in a later edit.

[82] If one pays attention some symptoms of this battle are hiding in plain sight, for instance how an article is often split across multiple pages for no real reason beyond gaining page views, or adds interspersed in a slide show of images.

Skype with an artist in London than it is to mail a postcard. But this is balanced against industry-to-person interaction also becoming more frequent; we are always in the line of advertiser's fire. hooks' "lively possibility where anything may flash itself on that screen" transforms again – *anything* is mediated by algorithmic Darwinism with code vying for the attention of code for the chance to pop up, and by the algorithm's portrait of the user drawn from constant observation, one's own habits defining a surprisingly small stable of potential content.

Rules (and consistently obeying them) are not the Terminator franchise's strong suit. A major example would be the explanation in the first film of the time travel machine by Kyle Reese (*future:* best friend and *past:* father of John Connor). Elucidating the caveats inherent to Skynet's machine, Reese says the machine is limited to only transporting biological things, thus he himself and the biological tissue-skinned Arnold Terminator but not clothing, and more importantly, not future weapons[1]. This clarification why the Terminator didn't bring a gun is insufficient: surely he could have had one wrapped in the close-enough-to-count false skin, or hiding somewhere beneath his own to be ripped out of an inconspicuous place (leg, chest etc.) upon arrival. And unsatisfactory at its initial purpose, this rule is then promptly broken when the completely non-biological T-1000 travels back through time with ease.

Perhaps what is most important in this type of fiction is not consistency but the illusion of consistency, at least in the eyes of an average viewer. All creative productions are a balance between intention and realization, concept and medium, with ambitions tempered by budget, technology, the limitations and expectations of an audience real or studio-imagined etc. In rules like the above restriction on what can travel in time, the internal world of the film and external world of film*making* collide; as long as this remains opaque to the average viewer any disjuncture between the two is immaterial.

As mentioned previously, the T-1000 has the ability to transform though this skill has limits. These limits are a mix of efforts at narrative consistency and a way to try to avoid what

---

[1] This may be also a rare example of exposition justifying *male* nudity in film, actors Michael Biehn and Arnold Schwarzenegger arriving in the past nude out of both narrative and prurient interests.

was to occur in the mind of viewers such as myself. Terminator 2 barely scratched the surface of what could be done if one were liquid metal, especially if one's imagination was allowed to run with the potent idea full steam ahead without the practical concerns of realizing these notions on film. What was the rubric under which the T-1000 operated? I'll let Schwarzenegger's Terminator and John Connor explain:

> John: I need a minute here, okay? You're telling me it can imitate anything it touches?
> Terminator: Anything it samples by physical contact.
> (John thinks about that, trying to grasp their opponent's parameters.)
> John: Like it could disguise itself as anything... a pack of cigarettes?
> Terminator: No. Only an object of equal size.
> (John's still reeling from meeting one Terminator, which now seems downright conventional next to the exotic new model.)
> John: Well, why doesn't become a bomb or something to get me?
> Terminator: It can't form complex machines. Guns and explosives have chemicals, moving parts. It doesn't work that way. But it can form solid metal shapes. [fig. 27]

This subtler break of the fourth wall by the imposition of film production limits on fiction is what we'll explore for a bit, going through the basic rules of the T-1000's transformation one by one.

It has to sample via touch whatever it imitates
By various commentators online this implies the need to touch the person and not just their clothing, causing a discrepancy when in one scene the T-1000 only touches a security guard's shoe and transforms into him. This to me is an improper 'gotcha' moment; it implies that sampling DNA

or some equivalent is somehow key to the machine's chameleon act, forgetting the T-1000's mimicry of not only a person's appearance, speech patterns etc. but also their clothing, sunglasses etc. This is a red herring; the true problem is the requirement of contact. This merely excuses the bare minimum of forms we see the machine take on.

The T-1000 could only be so small
This is reasonable; the T-1000 is a certain mass of material and though one could imagine it becoming bigger by incorporating some amount of a void within it but there is a minimum scale of it all. Key to this is a preference by the machine to remain whole, only splitting off due to extreme necessity or as the result of damage during conflict. Were it okay with being separated you could imagine it impersonating a child *and* a pile of books etc. and/or it becoming several small individuals/elements – throwing bits of itself like knives. Indeed pushing this to a limit-case: why not individual liquid metal particles leaping forth independently into the intended victim and reforming within them or some other such micro/nanotechnological murder method. Thinking back to self-reconfiguring modular robots: perhaps each cell or particle of liquid metal has intelligence but is only intelligent enough operating in conjunction with the whole, making the machine adverse to any splitting off of parts.

The T-1000 can't form complex machines, especially ones utilizing chemicals and/or moving parts
I'll excuse the chemical part, with the brief exception that its fairly easy to imagine a rifle or canon barrel being formed assuming the requisite accelerant and projectile could be acquired. But complex machines… certainly this is not limited by intellect. The complexity of the individuals and clothing the T-1000 imitates without much stress seems to imply that a few gears and levers shouldn't be out of the question.

Backtracking to the first condition, the low number of times the T-1000 transforms feels highly arbitrary in regards to the narrative and the character's maximum potential, but is likely far from arbitrary from a production standpoint as it allows both less special effects and a single main form/character for the audience to recognize as the enemy. Practically and strategically in a narrative sense, it would make far more sense for the T-1000 to change its form relatively regularly even if this for some reason would mean touching more people, a prerequisite easy enough to achieve for the fast and agile machine. This way, its target would have more trouble picking it out of a crowd and recognize it right away. At the very least, it could repeat the few forms it does inhabit. It would seem that the T-1000 which came to exist is hardly the ideal, suiting the narrative's needs at a bare minimum.

The transmissions from idea to text to screenplay & storyboard to feature film each invoke additions and subtractions which both suit the original ambitions and which temper them. In *Predator* (McTiernan, 1987) the changes between conception and production, and the resulting internal logic serves the plot instead of (or at the very least, in addition to) the financiers; with compromises made for the sake of making the performances feasible versus purely technical restraints. [fig. 28] A great illustration of the relationship between imagination and practicality is the iconic form of the titular Predator alien. The original character design had "a long neck, a dog-like head and a single eye", and "included 12-inch length extensions that gave the Predator a backward bent satyr-leg."[2] As the scenario required the Predator to be a very physical, active character and the movie was to be filmed in the jungle, this first try was naïve and unworkable. However the replacement design, far from being a cop-out, was very successful, perhaps more

---

[2] From the Wikipedia article on the character.

memorable than the specifics of the film itself, with crustacean-like mandible jaws (a suggestion of filmmaker James Cameron) hidden under an iconic mask and only revealed to great effect late in the film.

With the advent of more advanced motion capture technologies, this one-to-one of character and actor (the original *Terminator* and *Predator* being Schwarzenegger and Hall-sized, respectively) became only one solution to the creation of fantastic, roughly anthropomorphic creatures. Wide ranging scales and body types can now be portrayed by the same actor, typified by actor Andy Serkis in roles as varied as the small Gollum in the Lord of the Rings films, the massive title character in the recent *King Kong* (2005) and Caesar the chimpanzee in the recent *Rise of* and *Dawn of the Planet of the Apes* (2011,2014).

But at the time of the Predator such technology was still a decade away. In another inverse of the usual relationship between ideal and final result usually resulting in a subtraction or capitulation to pragmatism, the casting department and producers desired Jean-Claude Van Damme to portray the Predator due to a combination of his martial arts skills and his nascent star power. But when costumed and next to his large muscle-bound costars Arnold Schwarzenegger, Carl Weathers, and Jesse "The Body" Ventura a Van Damme-portrayed Predator didn't look threatening. He was replaced with the comparatively unknown 7'2" Kevin Peter Hall, an athlete-turned-actor trained in mime who went on to portray the Predator in the first and second films of the series. The resulting character was a mixture of compromise and surprising good fortune: while hardly as marketable by name, Hall's Predator was indeed imposing physically, and his acting training helped to bring to life a (mostly) silent creature, moving with an inescapable rhythmic ease towards his prey.

The Predators are an intelligent species, with a written and spoken language, and they can roughly ape human speech repeating this or that phrase they've heard in an approximation akin to a native English speaker echoing back a phrase they've just heard in French. The Predator shares the T-1000's skill with visual mimicry but instead of copying individuals, has a kind of light-bending adaptive camoflauge, mimicking on it's surface the world behind it to essentially become invisible. This technique/technology can be used either for defense or offense – camouflage can be used to hide, for instance a walking stick, a moth disguised against a tree, a skate hiding just beneath the sand on the seafloor. It can also be used to approach one's prey with impunity, think of a praying mantis, an anglerfish, an owl, a ghillie-suited sniper.[3] It add to its futuristic but still recognizable arsenal (nets, projectiles, lasers, a sort of sharp throwing disk, wolverine/Freddy Krugger-esque hand-claws and other analogue weaponry) the ability to see in multiple spectra including infra-red, giving it the ability to see the heat

---

[3] The Predator's camouflage technology has analogues (unsurprisingly) in current military research, with British defense contractor BAE Systems having developed a "system – called Adaptiv – using a matrix of hexagonal "pixels" that can change their temperature very rapidly." The system then scans the area behind the object to be hidden, say, a tank, and then displays that heat signature effectively cloaking the tank's heat signature in the infrared spectrum from the Predator (or heat-vision goggles or heat-seeking missiles etc.) And BAE Systems has also "combined the pixels with other technologies to provide camouflage in other parts of the electro-magnetic spectrum." Which is to say, projecting onto the skin of the vehicle (or eventually plane, warship, soldier) what lies behind it. Even more exotic, materials have been developed which slow, bend, and speed back up light around an object rendering it invisible. See: The Optical Society. "No Hogwarts invitation required: Invisibility cloaks move into the real-life classroom." ScienceDaily. April 30, 2015.

signature of a living creature – the Predator combines an ability to be less-visible with the ability to see more.

There is however a slight tell in the Predator's version of this kind of adaptive camouflage, an optical glitch or echo as it moves through the jungle. But where as the shaky state of the T-1000's liquid metal has to do with the limits of computer graphics technology at the time, the Predator's glitch manages to look visually stunning and to provide the otherwise overpowered creature some kind of limitation, necessary to create the sense, at least subliminally, that the Predator with all his advanced weaponry is still somehow vulnerable. And as a viewer the effect is entertaining - one could imagine the same sequences with perfect camouflage as less satisfying, the Predator rushing through the dense jungle illustrated with just shaking vines and the occasional ghost-like footprint.

This vulnerability is also accomplished by showing the Predator bleed. The Predator bleeds a day-glow green stuff – made as needed on set by mixing the liquid inside glow sticks and KY Jelly – which though odd is still recognizably blood, appearing as a result of injury. Thus the Predator's biology is made within the realm of the recognizable; it is a creature that can be harmed, cut, killed. Its blood also serves to highlight that what we are dealing with is certainly alien despite the general anthropomorphic form of the creature (as mentioned earlier, having forgone among other things the reverse jointed satyr-leg). Different colored blood is a classic troupe in films with aliens used to make them that much more 'other', for instance, his ears not enough, Spock's blood is green albeit a less dramatic shade.[4]

---

[4] From the IMDB page for the film: "The predator's blood - a goopy substance with the color of Mountain Dew - was made on-set using a mixture of the liquid from inside glow sticks, and KY jelly."

In order to defeat the Predator the last survivor of a group of special ops soldiers in the jungle for some manner of covert military operation but caught up in the Predator's deadly game of sport, Arnold Schwarzenegger's character Major 'Dutch' Schaeffer has to usurp the role of hunter from his pursuer. He utilizes a very analogue version of the same kind of blending into the background as the creature, lying in wait to ambush his opponent coated in mud hiding him both visually and, as importantly given the adversary's ability to see heat, thermally. It is not enough for him to be invisible in the sense we are familiar with, he must also appear inanimate to evade detection. Momentarily interred, Dutch is only able to ambush the Predator in concert with the terroir.

Movie critic Roger Ebert, usually an astute viewer, misses the core of what is going on in *Predator*. Though this detail is made a little clearer in the sequel, it is plenty apparent in the first film: the titular character is hunting for sport, mutilating and displaying his victims like trophies. Nevertheless Ebert found the motivations wanting, remarking:

> And the action moves so quickly that we overlook questions such as (1) Why would an alien species go to all the effort to send a creature to Earth, just so that it could swing from trees and skin American soldiers? Or, (2) Why would a creature so technologically advanced need to bother with hand-to-hand combat, when it could just zap Arnold with a ray gun?[5]

---

[5] Ebert, Roger. "'Predator' Sinks Its Teeth into Everything but Plot." *Chicago Sun-Times*, June 12, 1987. Ebert's review includes this great synopsis:

> The movie stars Arnold Schwarzenegger as the leader of a U.S. Army commando team that goes into the South American jungle on a political mission and ends up dueling with a killer from outer space. This is the kind of idea that is produced at the end of a 10-second brainstorming session, but if it's done well, who cares?

Equally dismissive, New York Times critic Janey Maslin quiped

> It looks like a man-sized lizard, can disguise itself like a chameleon, contains high-tech computer components and has dreadlocks on its head. Something for everyone.[6]

Terminator 2 eschews this potential to be misread by offering an overdose of voice-over and in-film exposition, yet viewed with some distance it is *Predator* that seems most logically consistent while the later feels stunted by technical and fiscal limitations. Neither as raw as the first *Terminator* nor as polished for a general audience as *Terminator 2*, *Predator* succeeds on its own terms. The film has aged well, enjoyable watched today without the Terminator films' cringe-worthy acting and pointless bombast. It feels both self-contained and presents much potential for further exploration – a species of highly advanced, intelligent, space faring creatures whose society revolves around hunting the most dangerous opponents they can find is a great counterpoint to humanity and our basically aimless expansion.

Like the Terminator series there are many sequels to the Predator. Additionally, two cross-over movies combining the Alien and Predator franchises, *Alien vs. Predator* and *Alien vs. Predator: Requiem* have been produced, this combo introduced as an idea in comic books and hinted at in a key scene inside

---

Ebert meant a ton to me, not because I agreed with him on every film but because his pure, child-like passion for movies was an inspiration. The art world within which I am entrenched lacks this kind of sincerity of investment in its criticality – voices that are at once individual and authoritative purely based on their commitment to the media.

[6] Maslin, Janet. "Schwarzenegger in 'Predator'" *New York Times*, June 12, 1987.

the Predator's space craft in *Predator 2*, where amongst a collection of trophies there is a recognizable Alien skull. The Alien creatures are violent, dangerous things; the suggestion that humans also qualify as worthy prey is a bit of a backhanded compliment – a nod to our own penchant for violence coupled with our ingenuity making up for what we lack physically.

Veering from where we've gone in this chapter, Predators, Aliens, and Terminators have a lot in common. Each represents a strange midway between science fiction and other genre forms such as horror and suspense; each is a fruitful franchise (quantity over quantity, anyway) far from done with producing sequels; each represents a creature from outside of our world (or time) presenting a difficult to deal with but not impossible to overcome existential threat, not just to the individuals we follow directly over the course of the narrative but at least in the case of Terminators and Aliens potentially to humanity as a whole. It would be worth considering why films featuring characters with so much in common would come into being and thrive around the same time, becoming franchises with no end in sight. Neither the mega rich miscreants of the (also endless) Bond films nor the lost and damaged souls committing murders in horror and suspense films, these characters approach humanity with goals (Predator: challenge for challenge's sake, Alien: propagation, Terminator: cold programmed purpose) separate from petty human ambitions. They confront the borders of what we call empathy (at least we, the viewers): well aware of these creatures' intentions we never the less fail to feel the least bit proud of their improvisations, their struggle. Instead, we sit on either side of an impenetrable border, we feel a powerful enmity towards these murderous creatures incapable of returning the distaste – so purpose bent, so alien.

When beginning to write this piece a decision has to be made: would I talk about the other Terminator films: *Terminator 3: Rise of the Machines* (2003), *Terminator: Salvation* (2009), *Terminator Genisys* (2015), and the television show *Terminator: The Sarah Connor Chronicles* (2008-2009)? I chose not to for a number of reasons. Firstly, the reason for beginning this project was my lingering interest in the T-1000 and the film where it first appeared. And I haven't seen the television series and didn't see the subsequent films in the theater (not watching them until relatively recently, and not enjoying them terribly much).

Financially the series peaked with *Terminator 2: Judgement Day*, descending from that films impressive 519 million dollar box office revenue to 433, 371, 322... Likewise the critical reception of the films has steadily declined, with the original, gritty film topping out at 100% approval on movie critique aggregator Rotten Tomatoes and 83 on similar site Metacritic down to (Rotten Tomatoes percent/Metacritic score) 93/75, 70/66, 33/49, 27/38. To an extent, the owners of the property agree with my aversion to the post *Judgement Day* follow-ups: the recently released fifth film pretends the third and fourth didn't happen.

Terminator 3: Rise of the Machines (Mostow, 2003) and *Terminator Salvation* (McG, 2009) are *retconned* out of *canon. Canon*, a word originating from the Greek *kanōn* meaning 'rule' which is meant in the context of fiction to refer to "the material accepted as officially part of the story in an individual universe of that story."[1] A word used originally in reference to biblical studies, critic Ronald Knox first used *canon* in reference to fiction in order to delineate between Arthur

---

[1] This and the following note about Sherlock Holmes cribbed in part from the wiki for "Canon (fiction)"

Conan Doyle's Sherlock Holmes stories and fan fiction and/or pastiche by other authors. [fig. 29] This is as opposed to (among other things) *fanon,* fan-made canon: facts about a particular fictional universe which reappear in multiple works of fan fiction or other non-canonical productions – "theories based on that material which, while they generally seem to be the "obvious" or "only" interpretation of canonical fact, are not actually part of the canon."[2] Sometimes fanon fills in blanks intentionally left open by the writers of a particular series to allow for future character growth, other times it appears in the form of small alterations and additions to a universe, such as the phrase "Beam me up, Scotty!" which never appears in the original Star Trek.[3]

Retcon, a portmanteau of *retroactive* and *continuity,* is a term I came across while compiling this book; some reading this will find it strange I hadn't heard of it before while others might encounter it here for the first time. It is not a terribly new gesture, but the word itself is quite recent; from the wiki on the term:

> The first published use of the phrase "retroactive continuity" is found in Elgin Frank Tupper's 1974 book *The Theology of Wolfhart Pannenberg,* 'Pannenberg's conception of retroactive continuity ultimately means that history flows fundamentally from the future into the past, that the future is not basically a product of the past.'

More commonly the term is applied to comic books. In comic universes multiple titles might engage the same character or

---

[2] From a guide to various television tropes tvtropes.org

[3] Equally interesting is 'ascended fanon', fanon which manages to slip into canon, such as (sticking with Star Trek) Uhura's first name Nyota, Sulu's first name Hikaru, and Captain Kirk being from Iowa, all ideas invented by fans which found themselves integrated into the series.

character set but not share a common continuity; writers assigned to a property might add or subtract events of previous books as needed to create the most advantageous narrative backdrop for the story being told. This may lead to a divergence between versions of a single character that feels too disjointed; such a character's narrative might eventually be *retconned:* writers will take the best elements of several streams of narrative and align them into a (semi) cohesive new timeline by adding predicate events and erasing others either through absence or correction.[4]

An early example would be Sherlock Holmes' apparent death in what was meant to be his Sir Arthur Conan Doyle's last Sherlock story, *The Final Problem*. It ends with Holmes having died along with his mortal enemy Moriarty:

> A few words may suffice to tell the little that remains. An examination by experts leaves little doubt that a personal contest between the two men ended, as it could hardly fail to end in such a situation, in their reeling over, locked in each other's arms. Any attempt at recovering the bodies was absolutely hopeless, and there, deep down in that dreadful cauldron of swirling water and seething foam, will lie for all time the most dangerous criminal and the foremost champion of the law of their generation.[5]

There is in this bit of text more than enough ambiguity (the bodies not being found etc.) as to leave hope for Holmes'

---

[4] The new official version will then be allowed to drift in multiple titles (the word for various series featuring perhaps the same main character such as for Batman the eponymous comic, 'Batman: Dark Knight', and 'Detective Comics'; for Spider-Man 'The Amazing Spider-Man', 'Spidey', 'Spider-Man 2099', 'Spider-Man')

[5] Doyle, Arthur Conan. *The Final Problem and Other Stories.* London: Penguin, 2011.

return, but this was not the author's intention. Doyle had felt the character had run his course and desired to be loosed of his creation's demand for new tales.

However, due to public displeasure about the hero's death — both the manner of his demise and in the implied end to new stories — pressure built upon the author to return to his star subject. Doyle first wrote a prequel of sorts which would turn out to be one of the most frequently dramatized stories, *The Hound of the Baskervilles*, and then resurrected the detective in the aptly titled collection of stories *The Return of Sherlock Holmes*. In *The Adventure of the Empty House* it conveniently turns out the hero had previously faked his death. Holmes is unimaginably smart, so this certainly falls well within the realm of possibility. But regardless of what could in hindsight be viewed as an intended twist, it is hard to read this as anything other than a capitulation to the desires of the reading public.

George Lucas is a famous user of the retcon. Where as most examples employee the act in a sequel or later work in a given universe, Lucas has shown a willingness to revise a finished, released film. In 1997 he started releasing 'special edition' versions of his Star Wars films in advance of the three prequels in production at the time. Multiple CGI-created aliens were added here or there in fitting with Lucas's stated original intentions. But more controversially, an early scene in (the fourth of the narrative but first produced giving the prequels their only somewhat deserved place in canon) *Star Wars: A New Hope* (Lucas, 1977) was altered. In it as originally scripted (despite claims to the contrary by Lucas) the up to that point morally ambiguous lead character Han Solo is having a conversation at gunpoint with a bounty hunter named Greedo intent on apprehending Solo and collection a reward. In the original version, Solo coyly aims and fires his weapon beneath the table to avoid capture. In the film it is somewhat unclear who shoots first and Lucas, desiring his character to operate without any sort of ambiguity of his heroic intentions, saw this scene as painting Solo in a negative light – a proper hero in his mind would never shoot unless shot at. Think of countless westerns – the type of film Lucas was raised on – in which the hero refuses to shoot the enemy unless the enemy reaches to draw his or her weapon. Through clever reediting and a little 'movie magic' Greedo shoots first in the revised, 1997 version, making Solo morally superior while leaving the mystery as to how at such close range anyone could miss their target unsettled.

Elsewhere in the series a more conventional retcon is performed. In *A New Hope* Obi-wan Kenobi tells Luke Skywalker about a character named 'Darth Vader' who killed his father. In *Empire Strikes Back* it is revealed that Vader was in fact Luke's father, but far from being a secret purposely foreshadowed by the first film, the intention was for the version laid out by Kenobi to indeed be the case, with Luke eventually avenging his father's death. The idea for Vader to actually be Skywalker's father came later during the writing of the sequel, and as it doesn't contradict the events of the first film its addition *seems* foreshadowed to anyone seeing the films in sequence.

Lucas's much maligned prequels have in the seventh film in the series, *The Force Awakens* (Abrams, 2015), themselves been retconned. For instance, a pithy pseudo-scientific explanation for the series's 'force' offered in episode 1 *The Phantom Menace* (Lucas, 1999) has been left by the wayside by director J. J. Abrams, tasked with making a good film while simultaneously staying true to the original trilogy and repairing the damage of the three prequels.

Wait, what just happened here? Begun as an interrogation into a very particular subject, the T-1000 and the Terminator series, all sorts of diversions into other things I want to talk about have snuck their way in. I've a limited number of analogies at my disposal, and interests, and methods of attacking a problem – these have all blended into a miasma of references from the hyper specific to the mundane. I'm halfway through this thing (as are you…) and I've resorted, if only briefly, to talking about Star Wars.

Author: Star Wars, I love you
Star Wars: *I know*

> In the long run, 500 years from now, everyone is going
> to be involved in some kind of information or
> entertainment. Nobody on the planet in 500 years will
> do a physically repetitive thing for a living.
> _-Venture capitalist Steve Jurvetson in "On the Edge of_
> _Automation" in MIT Technology Review_

> Henry Ford II (then CEO of Ford): Walter, how are you
> going to get these robots to pay union dues?
> Walter Reuther (head of the United Auto Workers
> union): Henry, how are you going to get them to buy
> your cars?
> _-An apocryphal conversation retold in 'Rise of the Robots:_
> _Technology and the Threat of a Jobless Future' by Martin Ford._

The Mechanical Turk is alive and well. The term comes from
'The Turk', a chess-playing automaton devised by Wolfgang
von Kempelen in 1770 consisting of a robotic puppet in
Turkish-dress behind a desk that could play a human
opponent chess. [fig. 30] It seemed a technical marvel; were
you to open some of its many cabinets you'd see elaborate
clockwork mechanisms, and while it operated you would even
here exaggerated whirring of parts. In truth the device hid a
skilled chess player hiding within the machine who could see
the game being played and controlled the Turk's play. The
level of sophistication of the contraption was astounding – for
instance the confederate operating the machine from within
would do so by candlelight, the smoke of which was ventilated
through a careful system terminating inconspicuously through
the Turk's turban. Napoleon Bonaparte and Benjamin

Franklin each played The Turk, and though eventually debunked in the 1820's the machine had a long career.[1]

The name lives on, sometimes used to describe hoaxes but generally the confidence game aspect of the narrative has faded leaving the term to describe human labor disguised as mechanized. Recently Amazon opened a Mechanical Turk market place in which Human Intelligence Tasks (or HITs) are farmed out to a 'global, on-demand, 24 x 7 workforce' – Amazon's Turk represents a system not of subterfuge hiding human-specific activities but rather a choice out of convenience and economics to have work done at a distance semi-anonymously. The trick in this case is the unforced but nevertheless hard to shake illusion that these 'Turks' neither fully mechanical nor human, and simultaneously that the humans poking there way through various tedious chores for as little as nickel a task are somehow more free, this piece-work a privilege afforded by technology.

Artist, art historian, critical theorist, and curator Michael Betancourt challenges the narrative of machines first coming for manual laborer:

---

[1] It is intriguing to think of this kind of hoax as having the potential to result in some good, enterprising engineers spurred on by a machine which seems beyond the realm of technical feasibility at the time (because it is). In attempting to recreate the phony devise they tackle problems that are as yet unsolved with a faulty confidence that the answer is known already and thus within their grasp. There are other examples of this in science, falsified research results in fields like particle physics and stem cell therapy leading to innovations simply out of efforts to reproduce a particular study or experiment. Or sometimes the systems developed to check a result themselves result in advances in the state of the art. Which isn't to excuse such a lie, simply to note the unintended consequences of even exaggerated progress.

In contrast to mechanized production's impact on the skilled trades, automation initially impacted *intellectual (immaterial) labor* rather than *physical (manual) labor* -- from the *Antikythera Mechanism* produced around 100 BCE, to the *Prague Orloj* clock from 1410, to the *Rathaus-Glockenspiel* in 1907 -- automation and automated systems have principally been concerned not with manual production, but with the elimination of intelligent labor.[2]

While this may be a bit of an exaggeration – simple machines such as the wheel, lever and inclined plane certainly started to replace manual labor if not with 'automation' than with 'mechanization' before the devices mentioned – it seems sensible to think of replacing humans performing HITs with machines and software as not simply something to-come but rather as the result of a process set in motion long ago.

The car service Uber is a great illustration of this process in which the replaced labor is both physical and intellectual. Uber is essentially a mechanical Turk: the user requests a bit of labor – a ride – from an anonymous cloud of providers via a phone and a car heads their way, symbolized by a small icon on a map which the user can watch as if watching a video game avatar travel across a level. There is a person behind the icon, a human – more often than not in my own experience Uber drivers have been more than willing to express their non-mechanized nature, telling stories about certain neighborhoods and their own path to this type of work and their experiences with it so far. However automated the dispatch system might be, driving is something reserved for humans, *for now*.

The original Mechanical Turk's job – chess – seemed the Human Intelligence Task par excellence, made the Turk's

[2] Betancourt, Michael. "Automated Labor: The 'New Aesthetic' and Immaterial Physicality." CTheory. February 5, 2013.

proficiency all the more miraculous-seeming to people at the time. But eventually real machines capable of performing this trick in the form of computer chess programs came into being. These softwares gradually grew in proficiency, first learning how to defeat and potentially train an average chess player and finally reaching maturity when the IBM supercomputer-run program Deep Blue rose in skill to the point it could defeat the strongest player in the world, the then world champion Gary Kasparov in a historic series of matches. [fig. 31] Their first clash in 1996 marked the first time a computer defeated a world champion in a single game in a tournament situation, though Kasparov won the match 4-2. The rematch in 1997 ended with Deep Blue the victor. Flash forward from that match in 1997 to now and computer programs runnable on a consumer laptop are more than capable of beating the best human player. In an exemplary match in 2014, the top chess engine at the time Stockfish ('chess engine' being the term for these programs, recalling Babbage's 'Difference Engine') defeated the then second ranked player in the world Hikaru Nakamura using both his personal expertise and a computer running the best program from six years prior. The highest live rating a human has ever achieved is 2889.5, held by current world champion Magnus Carlsen in April of 2014; Stockfish is sitting comfortably beyond that horizon at 3340. Furthermore, their human opponents far overmatched, chess engines are pitted against each other in competitions. These tournaments, though lacking in a casual fan base, reveal in the high level strategies employed by the softwares new ideas to employ by humans, if these strategies are even comprehensible to biotic contenders.

In what is largely considered to be an even more impressive feat, Alphabet company Deep Mind recently succeeded in building a system called AlphaGo which mastered the game of Go, a game with not only far more possible positions than chess, but also a type of game which resists a straight-forward evaluation of a given move or sequence of moves. It achieved this by trying to capture something akin to the intuition that a strong Go player depends on, using a neural network to examine 150,000 human games to try to learn what the most likely human moves would be and then training its search forward on these likely routes using its ability to evaluate the state of the board as a whole.[3] Additionally it performed the typical heavy lifting computers specialize in, running through far more permutations than humanly possible. Computer learning allows for an initial rough set of rules to be slowly edited, different rule sets playing simulated games against each other, slowly building within the system an ever more powerful Go player.

Which is all to say, HITs are *human* intelligence tasks until they aren't: until software and computation advance to the point that it is cost efficient (re: cheaper) to use machines and when machines aptitude for a specific task exceeds our own. A piece of software by researcher Babak Saleh and his team at Rutgers University combines powerful image processing technology to sort works of art by classemes, 'visual concepts' such as objects, actions, and style. The system then drew connections between different works which were cross-referenced with known art historical scholarship. In many instances, the connections it found, say, between Gustav Klimt and the Cubists, mirrored those already drawn by historians. But others were new, for instance "a link the algorithm makes between Frederic Bazille's *Studio 9 Rue de la Condamine* (1870) and Norman Rockwell's *Shuffleton's Barber Shop* (1950)."[4] The team notes "after browsing through many publications and websites, we concluded, to the best of our knowledge, that this comparison has not been made by an art

---

[3] For more about AlphaGo see Moyer, Christopher. "How Google's AlphaGo Beat a Go World Champion." The Atlantic. March 2, 2016. and https://deepmind.com/alpha-go.html

[4] Sokol, Zach. "An Intelligent Algorithm Made A Discovery That Slipped Past Art Historians For Years" The Creators Project. August 26, 2014.

historian before."[5] Correlations such as these can drive new paths for research, the machine spotting connections in datasets too vast to be mastered by even the most diligent historian. For example, one could imagine a new door into the process of Rockwell might have found by comparing classical artworks known to be circulated as mass-produced prints and major exhibitions hanging at the time and places Rockwell lived, with an eye for Bazille or other connections hit upon by software.

This type of example suggests we may stand to learn quite a bit about our selves at our most poetic and irrational (i.e. as when we make art) by allowing this type of observation from the outside to interrupt, reinterpret and augment our often myopic vision of who we are and what makes us human. Also it speaks to how the most human and abstract of research fields has a latent potential to be augmented by the algorithmic.

Referring back to Uber, venture captitalist Steve Jurvetson, a board member at SpaceX, Craig Venter's Synthetic Genomics, and Tesla Motors, was interviewed in a story subtitled with the provocative phrase

> Five hundred years from now, says venture capitalist Steve Jurvetson, less than 10 percent of people on the planet will be doing paid work. And next year?

He describes how in the near term we are experiencing a growth in jobs due to technology as the service industry becomes distributed digitally as in with Uber, Wag (a dogwalking service), Trashday (taking garbage cans from the side of the house out to the street), Helpling (housekeeping) etc. But Jurvetson considers each of these jobs as being on the edge of automation. As published, the following appears

---

[5] Ibid.

beside a jovial photo of Jurvetson, professionally lit before a black background:

> Everything about Uber has been automated except for the driver. The billing, the fetching—every part of it is a modern, information-centric company. Interestingly, what that means is as soon as automated vehicles arrive, that driver is easily removed. You don't have to restructure any part of that business.[6]

"When you look at manned taxis, 70 percent of the cost is actually related to labor costs," Hiroshi Nakajima, chief executive of the uncreatively-named Robot Taxi says, via a translator. "If we can replace that part with [artificial intelligence], I think we'll be able to provide a very attractive price point."[7] Robot Taxi intends to have a fleet of autonomous vehicles based upon existing car models ready for the 2020 Tokyo Olympics.

This replacement of the human workforce is a fate hovering in the future of many industries. Like Uber, much of Amazon's operation is already automated. They use an elaborate system of robots to automate more and more of their operations inside their vast warehouses, but the last phase of their operation, delivery from warehouse to consumer, relies upon outside companies like the US postal service and Fed Ex.[8] Amazon has been researching using

---

[6] Byrnes, Nanettte. "On the Edge of Automation." MIT Technology Review. September 28, 2015. It has to be noted how convenient/efficient this transition sounds... for Uber.

[7] Frommer, Dan. "Japan Is Building a "Robot Taxi" Service, with Thousands Planned for the 2020 Olympics." Quartz. November 2, 2015.

[8] Fed Ex (and other shipping companies) uses extensive operations research (OR) to determine the most effective routes, with a combination of software and human experience guiding routes etc. One could imagine their warehouses progressing in

drones to deliver their packages. From the "Amazon Prime Air" website FAQ:

> Q: Is this science fiction or is this real?
> A: It looks like science fiction, but it's real. One day, seeing Prime Air vehicles will be as normal as seeing mail trucks on the road.
> Q: When will I be able to choose Prime Air as a delivery option?
> A: We will deploy when and where we have the regulatory support needed to realize our vision. We're excited about this technology and one day using it to deliver packages to customers around the world in 30 minutes or less.

A whitepaper titled "Reinventing Retail: What Businesses Need to Know for 2015" included a poll which found an astonishing 88% of those polled would trust a drone to deliver a package.[9] Writer David Wagner (correctly in my opinion) describes this trust as being based on efficiency rather than a general fondness for autonomous delivery:

---

the same manner as those of Amazon, the vehicles becoming autonomous, with only the delivery from truck to door handled by a human until such time that a robot can handle that final step. Human labor will fall victim to something akin to the aforementioned notion of the god of the gaps: humans performing exactly the tasks not price-efficiently replaced by robots – until that price drops below that of a human workforce, the list of tasks that only a human can (efficiently) perform gradually declining.

[9] Wagner, David. "Drone Study Shows Consumers Are Ready." Information Week. February 19, 2015. In regards to the plurality comfortable with delivery drones, Wagner notes astutely notes

> You can't get 88% of people to trust scientists to believe the Earth goes around the Sun. Only 74% of people believe that. But 88% will accept a flying robot on their property, despite the fact that the No.1 user of drones to this point has been the military.

Consumers don't trust drones because they like flying robots buzzing their houses. They like drones because they like getting stuff fast. Solve that problem, and they won't care how you do it.[10]

The future of delivery by drone is never the less maligned by at the very least a vocal minority. And it has many regulatory and technical hurdles to overcome before becoming a commonplace reality, including mitigating/insuring against danger to humans. And not to be overlooked is the danger humans pose to the drones – the classic image of the pony express or train being robbed by masked men on horseback replaced with that of drones hunted by thieves aiming to shoot delivery-bots out of the sky with as little as a net or well-thrown rock. Regardless, these risks do not combine to make the idea a non-starter and research into this type of project continues, with initiatives from hard to dissuade companies like Walmart and Google in the works. The latter's 'Project Wing' hopes to have drone-based delivery up and running in 2017.

And beyond the dream of servicing the ravenous American consumer, drone delivery may find itself a reality faster elsewhere. Foster + Partners, a London-based architecture firm, is developing a series of 'droneports' in Rwanda. [fig. 32] Rwanda, nicknamed the 'country of a thousand hills' is relatively small but is hard to traverse with bad infrastructure susceptible to bad weather leaving isolated communities without necessities such as medicine.[11] In a place such as this,

---

[10] Ibid.

[11] Portions of this paragraph paraphrased from Webb, Jonathan. "Future Of Drone Delivery To Be Tested Out In Rwanda." Forbes. October 22, 2015.
There are of course questions. How do we manage a complex highway of air traffic? Who has legal liability in case of a crash? And how will we police those pesky drone bandits? [...] But these

drones present not only a novel convenience, but a dramatic improvement in living conditions, a potentially lifesaving tool.

And on the military side (if there's anything consistent across this narrative, there's always a military side), there's the Inbound, Controlled, Air-Releasable Unrecoverable Systems, or ICARUS for short. ICARUS is a proposed militarized version of robot delivery to provide support cargo to ground troops. But unlike the unintended failure of the mythical Icarus whose wax wings melted after he rose too close to the sun, ICARUS is meant to be unrecoverable, made of 'ephemeral materials' that can "sublimate from solid to gas and special glasses that shatter on command."[12] This disposable drone serves to buoy the sort of 'plausibly deniability' key to special operations activities which while always part of our military's operations form an increasing role as we scale back our large scale boots-on-the-ground military incursions.

ICARUS illustrates how the simple one-to-one replacement of human labor for robot oversimplifies the problem – ICARUS doesn't simply stand in for human piloted delivery, it adds features that no human could ever offer. A recent Onion article *Secretary Of Labor Assures Nation There Are Still Plenty of Jobs For Americans Willing To Outwork Robots* lays out the possible state of things to come for the modern worker:

> If you're available 24 hours a day, seven days a week, and able to repeat an extremely precise sequence of movements up to 150 million times without pause, there is work for you in this economy [...] I've heard from employers across the country, and they have assured me

are familiar teething problems for those daring to experiment. In the future, we may see dronemen top the continental superrich, as their fleet of robots swarm the skies.

[12] Axe, David. "Three Words: Self. Destructing. Drone." The Daily Beast. October 23, 2015.

these positions are up for grabs to any person with the ability to lift 250-ton payloads, bend steel plates into three-dimensional shapes, or withstand direct and continuous exposure to a variety of Level 4 biohazards.[13]

Advances in robotics have the potential to be a boon to human quality of living, a great example being advances in caring for the elderly with machines in the rapidly aging Japanese population. But robotics and automation are not quite a win-win. In an article for *The Atlantic* 'A World Without Work' author Derek Thompson suggests "The sanctity and preeminence of work lie at the heart of the country's politics, economics, and social interactions."[14] Thompson asks: "What might happen if work goes away?"

Capitalism breeds a strange contempt for any loss in maximum profits by those in power. Large management salaries are seen as a necessary evil but a raise to the minimum paid workers or to worker is seen as detrimental, a killer of innovation and of 'small' business. The latter is a sacred cow never quite quantified so as to make, say, raises to the lowest salaried of businesses too large to be considered 'small' also off limits. This has been seen quite vividly in the fast food industry. There has been pressure both from workers and from municipalities to raise the minimum wage, a wage at which much of the fast food industry's staffing hovers at or slightly above, but it is an industry in which cost cutting at every step in the process is seen as foundational to the business model. A prescient article in The Washington Post by Lydia DePillis's title frames this issue in a very specific

---

[13] "Secretary Of Labor Assures Nation There Still Plenty Of Jobs For Americans Willing To Outwork Robots." The Onion. September 8, 2015.
[14] Thompson, Derek. "A World Without Work." The Atlantic. June 22, 2015.

light. Its title sets the stakes: *Minimum-wage offensive could speed arrival of robot-powered restaurants.*[15]

Instead of robots as a future dystopian ruler, here we are given robotics as a different kind of threat: maintain the status quo, the gap between highly paid minority and low paid majority or else, in lieu of rebalancing the distribution of profits, robots will force the lowest paid contingent to lose their jobs. The article begins innocently enough with current McDonalds CEO Ed Rensi recalling the large, sub-1$-per-hour staff at a McDonalds in the 60's (when he began with the company) due to the amount of in-location preparation versus the long drive towards the current state of the majority of ingredients arriving pre-portioned and prepared.

The red herring is the correlation between increasing wages and labor-saving innovations: were wages to remain low, innovation would continue unchecked as there would still be money to be saved with one less hour of prep due to new cooking practices, a few more transactions per hour per cashier due to a new point-of-service system etc. The harvesting of raisins is a good example of this. Grapes and raisins are the third largest crop in California and the process requires a large amount of labor regardless of how that labor is priced or regulated and thus these work hours are always prime candidates for replacement.

> Retired USDA plant breeder David Ramming says he thinks he found the answer in a testing field of crossbred raisin grapes near Parlier. Ramming discovered Sunpreme in the mid-1990s, after crossbreeding raisin varieties that dry on the vine. The variety eliminates the need for workers to cut and hang grapes and the need

---

[15] DePillis, Lydia. "Minimum-wage Offensive Could Speed Arrival of Robot-powered Restaurants." Washington Post. August 16, 2015.

for paper trays for sun-drying. Ramming says this could be an answer to labor issues farmers face as the number of workers has decreased … "Basically, the raisin industry desires to be more like the almond industry – being completely mechanized," [USDA geneticist Craig] Ledbetter says. "Using Sunpreme, it is one step closer.[16]

Ramming's innovation fits a need not based on the increase in the price of labor hours, rather it is the difficulty in hiring the vast number of required workers needed due to a variety of issues including not only wage growth but the increased regulation of migrant workers from Mexico and Central Americal.

Rensi's argument lacks a direct correlation between a general drive to lower costs and the slowly rising minimum wage, ebbing upward while executive salaries have skyrocked, something not surprisingly unmentioned in Rensi's comments. The article seeks to tie the current drive to almost double the minimum wage – a move which only seems dramatic if you fail to account for the unaccounted for rise in the cost of living since the last major raise – to efforts to further automate things:

---

[16] Romero, Ezra David. "Sunpreme: The Grape That Could Revolutionize The Raisin Industry." NPR. October 7, 2015. Elsewhere in the article:

> Ron Kazarian farms raisins in Fowler, Calif., an agricultural community in the San Joaquin Valley. He says he likes the taste of the Sunpreme and is excited about its release, but he's afraid the raisins will fall off the vine too early.
> This variety is falling down on the ground before the machine even gets in the row," says Kazarain. "What do you think is going to happen when the vine begins to shake? We would be concerned that all the fruit would be on the ground.
> He plans to solve that problem by creating a mechanical harvester that picks the crop by shaking the vine and vacuuming raisins off the ground.

burger-flipping robots — or at least super-fast ovens that expedite the process — become that much more cost-competitive if the current federal minimum wage of $7.25 an hour is doubled.[17]

But even though there is some sense in this assumption that a higher cost of labor leads to higher pressure to mitigate those costs, the thin veneer hiding a general focus on profit before everything shows. David Brewer, chief operating officer of a kitchen equipment company (who certainly stands to gain from a race to automation:

> The miracle is, the wage increase is driving the interest," Brewer said. "But the innovation and the automation, they're going after it even before the wages go up. Why wait?"[18]

"Why wait" indeed. Olive Garden is in the early stages of considering replacing much of the job of its waiters with 'Ziosk', a tablet-based ordering system leaving servers only the task of delivering the food, a task itself almost too easy to imagine being replaced by robots once customers are made comfortable with the idea. Robots are used as a threat to pit anyone not at the bottom against those who will gain from a minimum wage increase:

> "Even though sushi chefs tend to make more than $15 an hour, they could be on the chopping block if servers need to make $15 an hour, too."[19]

Persona Pizzeria's vice president Harold Miller suggests

---

[17] Ibid
[18] Ibid
[19] Ibid

all the automation working its way into restaurants could eventually cut staffing levels in half. The remaining employees would just need to learn how to operate the machines and fix things when they break.[20]

Sure, until some manner of maintenance bots begin doing house calls, roving from business to business maintaining the robotic staff. The article spends its last third giving voice to those who think restaurant goers will miss the 'human touch'

> Hudson Riehle, the National Restaurant Association's senior vice president for research and knowledge. If they're not careful, restaurants could jettison the one thing that kept people coming through the doors.[21]

But we should question the point of this type of piece. It would be a different story if there were examples of companies which, having struck a state of profitability, stayed making the amount of money they were making with management and ownership (whether privately owned or in the form of stockholders) happy with the status quo.

Capitalism demands every increasing productivity, market and/or market share growth, and whether this is accomplished via product innovation, an increase in consumer interest or increases in productivity is beside the point. Robotics both as reality and as threat serve as one of many ways to diminish the expectations of the working class, happy with what they get 'or else'.

And for those post-fordist workers typing away, seemingly immune from the automation of labor, softwares lie just beyond the horizon growing in capacity and preparing to swoop in. Low-level tasks formerly performed by humans are

---

[20] Ibid
[21] Ibid

being tackled by machines at a growing clip. Software that can assist with the time-intensive operation of legal discovery, in which one party in a lawsuit is allowed access to the evidence in the hands of the other party is beginning to see usage. And sports journalism at the level of game summaries and headline writing is becoming automated. A study from the University of Oxford ranked occupations based on the likelihood of their usurpation by machines:

> Choreographers, elementary school teachers, and psychiatric social workers are probably safe, according to that analysis, while telemarketers and tax preparers are more likely to be replaced.[22]

Jurvetson concurs:

> What you're farming out to humans today are those things that computers just barely can't do. We know from Moore's Law and improvements in computing that in two or three years [much of this] work will be automated.

> If a startup or new business venture has created a job that involves human labor, it probably has done so in a way that is pretty marginal. Whether you're a technology enthusiast or a detractor, the rate at which this will shift is probably going to be unprecedented. There will be massive dislocation.[23]

Jurvetson is far from apologetic, describing a future where we "live off the production of robots free to be the next Aristotle or Plato or Newton" (apparently we have a very specific, high-minded set of interests in this future), and noting the

---

[22] Byrnes, Nanettte. "Work in Transition." MIT Technology Review. September 28, 2015. The previous few sentences are paraphrased from this article.
[23] Byrnes, Nanettte. "On the Edge of Automation."

inability of the political process to deal adequately with the dramatic shift he sees as inevitable: "I see zero chance that long-term thinking will govern policy."[24] The role of venture capital is to, by being ahead of the curve versus traditional investment, access a field with much higher risks and much higher rewards. The latter drives one's practice, not the relationship between those rewards and the general wellbeing of society. But the added abstraction of capital aides in distancing the two types of consequences, moral and fiduciary. Jurvetson and his blunt honesty in this article is a valuable window into a system and its profit-maximizing tendencies with their resulting consequences for the future.

There is research happening to understand what it will be like to work hand-in-hand with machines, with robots designed to share the workplace with humans. But its hard not to feel that such a project is inherently disingenuous (even if not intentionally so) as eventually those humans who do find themselves comfortable with robotic assistance may as well accept that they are essentially training their replacements – the most embittering of ignominious tasks expected of the soon to be unemployed.

Indeed, as the title of an article in New Scientist puts it, *The hot new job in Silicon Valley is being a robot's assistant.*[25] Much in the same way we might pull out a calculator for the tip at a restaurant, some new softwares which incorporate machine learning are also adding human input to the mix. What Facebook terms 'human trainers' sit behind the scenes of M, their new AI-based digital assistant in their mobile messenger software currently being beta tested. The article's author gives several examples of this type of human-machine partnership, including Interactions, a Massachusetts based developer of

---

[24] Ibid

[25] Rutkin, Aviva. "The Hot New Job in Silicon Valley Is Being a Robot's Assistant." New Scientist. November 3, 2015.

"digital conversation assistants." The author gave Hyatt's reservations line (which uses Interactions to field the millions of calls they get each year) a ring:

> A pleasant male voice answered. It sounded human, but too crisp to be a live operator. "How can I help you?"
>
> "I'd like to book a room," I said. I wanted to make my voice too hard for the AI to parse, so I tried to mumble when I spoke. Spanish pop music blared through cafe speakers behind me.
>
> Then there was a pause. It lasted maybe three seconds, the amount of time I might attribute to someone studying a sheet of paper or typing my request into a computer, if I'd been on the phone with an ordinary person. Was this my moment of human contact? Was anyone listening?[26]

One can't help but feel we are being fooled, with researchers mixing mechanical and technical skill set accumulation (picking a selection of parts from bins to be taken to a place for assembly for example) with the language and social skills that allow a subsequent stage of full human-replacement to occur smoothly.[27]

---

[26] Ibid.

[27] Melonee Wise, founder of Fetch Robotics in San Jose, describes well meaning efforts to make working with machines easier for humans:

> When a robot talk, it conveys a certain level of intelligence, and people start thinking the robot is smarter than them, so they're less likely to help the robot," she says. "When the robot has nonverbal cues, people are much more willing to help out.

In Knight, Will. "Several Startups Are Selling Robot Workmates." MIT Technology Review. September 28, 2015.

Is there a battle being fought for wages between human laborers and that of machines? Or is this a proxy war hiding the real animosity between capital and individual lives? Some competitions are in the open – imagine a large field, opposing forces lined up and then at once the two sides head towards each other in a violent convergence. Others happen behind the scenes, purported cold wars in which technological advances from adversaries push forward science on both sides, so-called 'cold' because whereas proxy wars are sparked and armed by the major combatants these main opponents themselves remain unharmed, able to hold a chess match in plain sight with pieces on the board the lives and livelihoods of strangers. The biological and capitalogical worlds pit human workers and machines against one another; the latter's horse in the race does its best to claim territory by convincing us it is a)undesirable and b)better off managed by its systems. Work is something for machines and for code; what it is we are to do in the wake of the end of wage labor to acquire an income is left unspoken.

Now for a bit of a detour: biologist L. Van Valen in 1973 proposed an evolutionary theory called the Red Queen Hypothesis. The name derives from the character in Lewis Carroll's *Through the Looking Glass*. Alice finds herself with the Red Queen in 'The Garden of Live Flowers'

> 'Now! Now!' cried the Queen. 'Faster! Faster!' And they went so fast that at last they seemed to skim through the air, hardly touching the ground with their feet, till suddenly, just as Alice was getting quite exhausted, they stopped, and she found herself sitting on the ground, breathless and giddy.

> The Queen propped her up against a tree, and said kindly, 'You may rest a little now.'

Alice looked round her in great surprise. 'Why, I do believe we've been under this tree the whole time! Everything's just as it was!'

'Of course it is,' said the Queen, 'what would you have it?'

'Well, in our country,' said Alice, still panting a little, 'you'd generally get to somewhere else—if you ran very fast for a long time, as we've been doing.'

'A slow sort of country!' said the Queen. 'Now, here, you see, it takes all the running you can do, to keep in the same place. If you want to get somewhere else, you must run at least twice as fast as that!'[28] [fig. 33]

Van Valen suggests a kind of evolutionary (arms) race. In order for a species to survive it must always keep moving, evolving. For instance, predator and prey are constantly evolving to be respectively better hunters and less-vulnerable food, with the penalty for either side stopping in this process being extinction. This pressure occurs not only vertically along the food chain, but also horizontally, with each creature evolving to compete for a finite number of local resources.

This process of coevolution requires a system whereby quick mutations are possible and rapidly transferable – sexual reproduction provides exactly such a system. Sexual reproduction is tailored to keep more complex animals one step ahead of their malignant microscopic counterparts such as bacteria and other parasites; instead of a slow drift of mutations in a cell line, two distinct genetic make-ups are intertwined for every subsequent generation. A side effect is a

---

[28] Carroll, Lewis. Through the Looking-Glass and What Alice Found There. Macmillan, 1871.

slow accumulation of bits of genetic data as bits of DNA find themselves not expressed in the offspring, a patchwork of sometimes harmful, sometimes beneficial modifications lying dormant in the genome.

A great example of this in action is the harmful mutation that causes sickle cell anemia. A disease in which the affected individual's blood cells develop malformed and less capable of transporting oxygen. If a child is passed a single copy of the gene it lays dormant but if two copies are shared the child suffers from the disorder. Why does such a potentially detrimental piece of erroneous code survive in our genome? Left to our own devices it might find itself eradicated, but we are never on our own. Humans live amongst a complex web of interspecies competition, and hidden in our genome are various weapons for dealing with the more malignant cohabitants. Sickle cell is most prevalent in Africa and among those elsewhere of African decent – it turns out that having a single instance of the sickle cell gene makes one less vulnerable to the symptoms of malaria.

This mutation with its pros (making malaria less deadly) and cons (sickle cell anemia to those acquiring a pair of the mutated genes) has

> occurred independently, at least four times in human history, with the same gene involved, indicating that this gene must have been preserved for some reason and parasitic pressure caused it to be manifested.[29]

The DNA of each species is littered with such innovations with new ones sputtering into and out of being constantly – should a crisis arise in which a parasite evolves to be successful to an extent that a species is faced with the pressure

---

[29] Shuttleworth, Martyn. "Red Queen Hypothesis - The Evolutionary Arms Race." Explorable.

of extinction, these genes act like genetic safe rooms into which enough who are immune or less-susceptible can weather the storm. Predators and prey, parasites and hosts, but also cohabitating trees etc. are all pressured to keep running, to keep trying new codings while maintaining a stockpile of tools in case of emergency. Both the best of a species for the general case, present situation *and* the sub-optimal elements which maybe come in handy once an epoch are useful.

Returning to our subject, The Red Queen Hypothesis suggests the likely way forward for our species in concert with the mélange of individual robots and AI: we will compete for wages, for relevance. Each will increasingly experience the other as a purely maintenance system, us maintaining the machines where they can't do it for themselves and they caring for ever more of our simple needs. Each has vestigial-seeming traits that may seem irrelevant in the current-world but may spring back into consequence. And should either side cease developing in concert with the other, crises will emerge. Think of the twin fictional poles of AI-induced disaster and the much rarer image at the end of Dune of AI finding itself outlawed, then imagine the infinite subtle variations between these extremes.

The labor market is a field where early skirmishes between human and machine will play out. As human tasks both physical and mental fall under the increasing purview of machines intelligent and otherwise, we will feel the pressure to mimic the automaton. Much has been made of the transition from physical to cognitive labor, post-fordism etc. This transformation creates a new type of ideal worker, always on call via smart phone, with a soft skill set moldable to new tasks and techniques – skilled labor becomes meta as the skill to acquire skills supersedes specialization. This changes the rules of the game for exploitation. Think of a factory worker, a type of work which still goes on; we wear as much clothing and drive as many cars as we always have which is to say these things are still for the most part made a minimum of human assistance. A worker may sit in place for hours, attaching one part to a complex toy. Their work neither is sellable on its own outside of the whole chain of production nor is it specific to the individual – should the worker be injured another unskilled worker would quickly fill their place in the line. These jobs offer a simple wage for labor relationship – it is hard to imagine the assembly line worker wishing to make one more doll, work one more minute without compensation.

The new post-fordist worker on the other hand gladly offers up her own time for the sake of advancing in their career – think of someone at their laptop on a train, busily coding away on the way to and/or from doing the same thing at a desk.[30] AI and humans will evolve to become ever more like each other, what value remains for those cognitive tasks still remaining exclusive to humans will be ever more tied to a computer-like 24/7 work cycle. Perhaps survival lies in a regressive trait-like tendencies of humans: laziness, stubbornness, a willful need to rebel against authority, our mockable instinct to sneak off mentally from the task at hand to communicate or be communicated at via social media etc. While each of these under optimum circumstances might be a drag on both productivity and general happiness (other, harder workers unjustly bearing an unfair amount of the burden), as the stakes are continually raised in the battle for labor between humans and machines it may be this type of gestures which hold a potential key to riding out the storm.

------

[30] These conversations about the state of these new economy laborers are important, but we'd be wise not to forget that to focus solely on these relatively well-compensated, more educated workers would be to ignore the vast amount of wage labor still performed on a daily basis. Fast food, factory work, mining, farming etc. continue to be the majority of the activities done by Americans, and even more so in less developed nations.

Mental versus physical labor... we'll end this shallow wade into a subject due a lot more research with another bit from the same Onion article referenced at the beginning:

> If you're unable to outperform a robot as a manual laborer, there are plenty of jobs that allow you to make a living with your brain rather than your hands," Perez said. "As long as you can sit at a desk and complete in excess of 34 quadrillion calculations per second in your head, I guarantee you will be able to find work.[31]

---

[31] From the same Onion article.

> Mr. Ephraham: The T-Portable has four layers of cross-referenced sensors. It is by far the safest car on the road today. No one has ever died or been injured in any car accident caused by my T-Port.
> Lucca (lawyer): That would be news to my client.
> Mr. Ephraham: That accident was not caused by my car. It was caused by one of my employees, who took the car for an unauthorized test drive. [...] Ms. Searle's injury is unfortunate, but it has nothing whatever to do with my car. Technology can overcome most obstacles, but it can't overcome human nature.
> -*The Good Wife, Season 7 Episode 7 'Driven'*

Fully autonomous weapons, systems which make the decision to pull the trigger without human intervention are a subject of discussion in the United Nations, with an effort to ban them pre-emptively begun but unlikely to succeed. Noel Sharkey, a professor of artificial intelligence and robotics at the University of Sheffield and co-founder of the International Committee for Robot Arms Control, a coalition of robotics experts who are campaigning against the military use of robots, describes in no uncertain terms the opinion that

> We shouldn't delegate the decision to kill to a machine full stop. Having met with the UN, the Red Cross, roboticists and many groups across civil society, we have a general agreement that these weapons could not comply with the laws of war. There is a problem with them being able to discriminate between civilian and military targets, there is no software that can do that.[32]

---

[32] Grant, Harriet. "UN Delay Could Open Door to Robot Wars, Say Experts." The Guardian. October 6, 2015.

Disregarding with the well-documented history of humans lacking the ability to avoid civilian casualties and/or the potential for machines to at least improve upon our shabby record of accidental bombings etc., it seems reasonable to worry about artificially intelligent weapon systems. But in an article responding to recent discussion around the potential dangers of artificial intelligence-driven weapons, authors Zach Musgrave and Bryan W. Roberts see the problem differently: *Humans, Not Robots, Are the Real Reason Artificial Intelligence Is Scary.*[33]

The Musgrave and Roberts article is in response to *Autonomous Weapons: an Open Letter from AI & Robotics Researchers,* an open letter and petition on the website 'Future of Life' which seeks to forestall the development of autonomous weapons, defined as those which select and engage targets without human intervention versus semi-automated devices such as cruise missiles which operate at the direct bequest of human operators. The petition goes on to say

> Autonomous weapons are ideal for tasks such as assassinations, destabilizing nations, subduing populations and selectively killing a particular ethnic group. [...] Starting a military AI arms race is a bad idea, and should be prevented by a ban on offensive autonomous weapons beyond meaningful human control.[34]

The petition has many esteemed signatories including Stephen Hawking, Elon Musk, Steve Wozniak and Noam Chomsky. But is the effort too late? A predecessor to this type of weapon that can 'select and engage targets without human intervention' has already been deployed. Designed to combat

---

[33] Musgrave, Zach, and Bryan W. Roberts. "Humans, Not Robots, Are the Real Reason Artificial Intelligence Is Scary." The Atlantic. August 14, 2015.
[34] Ibid.

the destruction of a delicate coral reef by the crown-of-thorns starfish *(Acanthaster planci)*, the COTSBot designed by the Queensland University of Technology uses relatively simple software and machine learning to target these starfish, locating them and injecting them with a deadly poison. [fig. 34] The COTSBots may seem innocuous enough as they attempt to defend a much loved endangered ecosystem but they are frightening for many reasons, not the least of which the way these bots recall fictional evil robots such as some of the many machines featured in Michael Crichton's directorial debut *Runaway* (Crichton, 1984), small poison needle-wielding robots built by a mad scientist. COTSBots and like machines granted the power to locate, sentence, and exterminate considered-lesser species free of human intervention may be the first steps down a slippery slope between machines executing lower lifeforms, machines killing invasive mammal species, machines enabled to hunt and kill groups of humans considered lower – one need only to look at the type of inhumane treatment afforded terrorists for a glimpse at our ability to draft a different set of rules regarding our fellow man, "All animals are equal *but some animals are more equal than others*"

Musgrave and Roberts also signed on to the Future of Life petition, however their position differs slightly. For them the weaponization of weak-AI enabled devices poses a greater threat and will proceed any danger posed by a future self-aware system or a state-sponsored strong AI arms race. Musgrave and Roberts conclude their argument with a great basic question: "What counts as intelligence, and what counts as a weapon?"[35] Strong AI remains firmly beyond the horizon; Musgrave and Roberts reference a recent (supposedly) Turing Test-defeating chatbot named 'Eugene Goostman' being stymied right out the gate by MIT professor

---

[35] Musgrave, Roberts. "Humans, Not Robots, Are the Real Reason Artificial Intelligence Is Scary."

Scott Aaronson. At the risk of loosing our train of thought, what follows is an excerpt from the conversation between Aaronson and 'Goostman' from Aaronson's website:

> Scott: How many legs does a camel have?
> Eugene: Something between 2 and 4. Maybe, three? :-)))
> By the way, I still don't know your specialty – or, possibly, I've missed it?
> Scott: Do you understand why I'm asking such basic questions? Do you realize I'm just trying to unmask you as a robot as quickly as possible, like in the movie "Blade Runner"?
> Eugene: …wait
> Scott: Do you think Alan Turing, brilliant though he was, had trouble imagining that the judges of his "imitation game" wouldn't think to ask commonsense questions like the ones above—or that, if they did, they'd actually accept evasion or irrelevant banter as answers?
> Eugene: No, not really. I don't think alan turing brilliant although this guy was had trouble imagining that the judges of his imitation game would not consider to Oooh. Anything else?
> Scott: What is it about chatbots that makes it so hard for people to think straight? Is the urge to pontificate about our robot-ruled future so overwhelming, that people literally can't see the unimpressiveness of what's right in front of them?
> Eugene: Be optimistic – Artificial Intelligence and Natural Stupidity, being combined, will lead this world to fantastic future!

What counts as *intelligence*, and what counts as a *weapon*? Perhaps we've over discussed the former a bit, lets discuss the latter a bit. First, a case study: is a car a weapon?

A car can certainly be used to kill. In the summer of 2003, an 86-year-old man named George Weller drove his large Buick through the length of the crowded Santa Monica Farmers' Market, reaching somewhere between 40 and 60 miles per hour, killing 10 people and injuring 63.[36] By the time it was over,

> The hood of the 11-year-old Buick was mangled and dented. An apple core and two unmatched women's shoes lay atop its roof. "It was gruesome," a witness said "There was fruit everywhere, and [bodies] were covered with raspberries and other things."[37]

Charged with manslaughter, Weller pleaded not guilty but was found guilty on all charges; his advanced age kept him from prison. Witnesses disputed his claim that it was purely an accident, noting he looked straight ahead as bodies flew past his windshield and didn't try to brake. But giving Weller the benefit of the doubt, perhaps this story is more like a gun accidentally going off than one being brazenly fired into a crowd.

Things get a bit darker in the case of so called hit-to-kill accidents in China. In these cases (many of which are fully documented) a driver accidentally hits a pedestrian, injuring them. The driver then proceeds to back up over the injured person multiple times to kill them, as the fine for killing a

---

[36] The event inspired a Southpark episode titled 'Grey Dawn' in which senior citizens start indiscriminately running over the town's citizens. Indeed this scenario feels more and more plausible as the percentage of elderly drivers increases year after year.

[37] Rubin, Joel, Daren Briscoe, and Mitchell Landsberg. "Car Plows Through Crowd in Santa Monica, Killing 9." *LA Times*, July 17, 2003.

person is less than $50,000, far cheaper than the potential cost of caring for an injured victim.[38]

Lets up the ante from manslaughter and murder as economic expedient to full-blown murder:

> A seven-year-old boy is reportedly among the three people killed in Austria by a man who ploughed his car into crowds in the country's second-largest city and then reportedly started stabbing people. […] A statement from the city council said: "At 12pm there was an appalling incident in the centre of Graz, which has caused major alarm and left the city deeply shaken. A killer used his car as a weapon and deliberately ran people down on a rampage. The perpetrator is in custody."[39]

There is also the Hebei tractor rampage in China 2010 in which an inebriated factory worker drove a bucket-loader first over his boss, then into innocent bystanders (including a five-year-old), killing 17 and wounding 20. And a few years previous in Japan there as the Akihabara massacre, in which a man drove a truck into a busy shopping area, killing three and injuring two (after which he proceeded to stab to death another four – the Austrian attacker also adding stabbings to his repertoire).

This is round about way to get to Musgrave and Roberts' point: "Unlike self-aware computer networks, self-driving cars

---

[38] See the very dark "Sant, Geoffrey. "Driven to Kill : Why Drivers in China Intentionally Kill the Pedestrians They Hit." Slate. September 4, 2015.

[39] Dearden, Lizzie. "At Least Three Killed in Austria after Man Drives into Crowd before 'stabbing Passers-by' in Graz." *The Independent*, June 15, 2015.

with machine guns are possible right now."[40] And forget the machine gun, the previous examples show that the car on its own can do plenty of damage. Some of the potential for autonomous automobiles to cause harm became the subject of an episode of the television show *The Good Wife*. *The Good Wife* has an overarching narrative stretching from episode to episode involving a former lawyer who stands by her disgraced, unfaithful husband and takes back up her law practice. But each individual episode plays out a simultaneous legal drama often revolving around subject matter 'torn from the headlines.' In the case of season seven episode 'Driven', the subplot revolves around an accident involving a self-driving car called the T-Port and a human driver, the human having sued the T-Port's maker. One of the scientists behind the car's software, a Mr. Dudewitz, gives a definition in which quickly devolves into territory we've covered a bit already, but is a bit too good to leave out:

> Lawyer: One last question. How do you think the car's hard drive got erased?
> Do you think that the car's... capable of erasing its own memory?
> Dudewitz: Yes, I do.
> Lawyer: And if you believe that, don't you think that the software could override the safety features?
> Dudewitz: I don't think I would go that far.
> Lawyer: But if you believe that A.I. can transcend its creators' original mission, and you introduced the theory of fuzzy driving in order to make the car more aggressive, more human-like, then why couldn't the computer override your safety controls?
> Dudewitz: I am a smart man. A genius. Yes, and I've created something so complex, I can't positively tell what it can do and what it can't.

---

[40] Musgrave, Zach, and Bryan W. Roberts. "Humans, Not Robots, Are the Real Reason Artificial Intelligence Is Scary."

Lawyer: So it could have put that woman into a wheelchair?
Dudewitz: It's possible.[41]

But the potential to pin the accident on the car's sentience is short lived as it turns out that individuals at the company had hacked the car. As described by one of the lawyers,

> That's right. Humans. Always there to disappoint. They wanted to spook him, but they crashed his car into your car, ma'am, I'm sorry to say.

It is the human element that is truly frightening about autonomous or semi-autonomous cars. One can imagine a motivated terrorist simply modifying a few (soon to be) off the shelf autonomous vehicles and setting them loose, whether directed at a specific target or into a random crowd. The machine gun element in Musgrave and Roberts' piece is a nice touch, but the disruption that would be caused by several non-armed cars coming to life as in Stephen King's *Christine* and attacking pedestrians would be hard to stomach, immediately disrupting the whole field of autonomous vehicles (much as 'weaponized' aircraft did in 2001). Furthermore, think of a group of Uber cars, self-driving like the Jonny Cabs in *Total Recall* (Verhoeven, 1990), suddenly

---

[41] More from this exchange:
 Lawyer: Are we in danger from A.I.?
 Dudewitz: I think it's getting smarter. Every day. Learning our boundaries and its boundaries... it's evolving. I think there will be an adjustment period after it takes over, but eventually...
 Lawyer: Wait, I'm sorry to interrupt. Um, what do you mean by "takes over"?
 Dudewitz: The singularity. When the system is capable of recursive self-improvement, when it is better at recalibrating, expanding and spawning than we are, the brief blip of humanity's reign will reach its inevitable conclusion.

taken over by someone with malicious intent and veering into oncoming traffic.[42] This is not to mention the same threat as applied to automated aircraft. If we've learned anything this millennium it is that we needn't make a huge technological advance to create new things to fear. One need only take a few givens – the mass and speed of an airliner and the natural aversion to being cut by a boxcutter or any other sharp implement – and combine them, mining a small gap in security procedures. What autonomous vehicles add to the equation is a subtraction: in the previous examples of driver-made mayhem, the drivers either die before the end of the day's proceedings or end up in jail; in the alluded to 9/11 example the attackers give up their own lives; in a more recent attack involving a bomb hidden in a laptop only the bomber died. With autonomous or semi-autonomous vehicles this price in human life for the attacker is removed or at the very least forestalled (should the culprit eventually be apprehended), leaving only as barriers the will and skill to carry out an attack.

Cambridge professor Martin Rees has co-founded at Centre for the Study of Existential Risk (along with Huw Price of Cambridge and Jaan Tallinn, co-founder of Skype). Rees wants the discussions to begin around the dangers up ahead; he described in an editorial for Science,

---

[42] Raymundo, Oscar. "Uber's First Autonomous Car Goes out for a Test Drive." Macworld. May 22, 2015. More:

> If everything goes according to Uber's master plan, pretty soon you won't have to sit through having to make small talk with a human when you take one of their rides. The ride-sharing giant has set up a research and development outpost in Pittsburgh, and its first project has already driven itself off the lot for a test drive. Uber now has an autonomous car.

True, it is hard to quantify the potential "existential" threats from (for instance) bio- or cybertechnology, from artificial intelligence, or from runaway climatic catastrophes. But we should at least start figuring out what can be left in the sci-fi bin (for now) and what has moved beyond the imaginary.[43]

I think the terminology is a bit of an overstatement. There may be larger existential risks to humanity as a whole that if left unchecked may result in extinction. For the most part, however real a given risk such as global warming may be, this most have only a slight chance of wiping us out as a species – we are a resistant lot, surviving (if not thriving) in the many of the most extreme environments on earth. Often what are described as existential risks have more to do with being risks to a certain minimal quality of life and avoidance of the catastrophic than they do extinction. There are real, ultimate dangers such as all-out nuclear warfare and there are relatively small threats, with as their limit case September 11's death toll both immediately and as compounded by the subsequent wars it was used as grounds for. The death of a thousand or a million or even a billion people would be a terrible loss, but any attack or war – barring a thorough nuclear assault on all sides which manages to reach the farthest reaches of the world with its poisonous radiation – is not *existential* in its threat.[44] Existential risk seems to be the wrong way to phrase the threat, though we should recognize the potential harm up-to-but-not-including our erasure.

What counts as intelligence? We talked a little bit earlier about the difficulties of distinguishing (without resorting to semantics) between sufficiently advanced AI and human

---

[43] Rees, Martin. "Denial of Catastrophic Risks." *Science*, March 8, 2013.

[44] See the brief addendum about nuclear weapons at the end of this volume.

intelligence, but there are still barriers to becoming 'sufficiently advanced.' Lets delve again into that abyss.

> When the seductive charms of fiction penetrate reality they arouse first astonishment, then misgiving, and finally a dread of man's being dispossessed by his own inventions.
> *-Marc Auge in The Future* [45]

> I for one welcome our new computer overloads.
> *-A parenthetical to former-Jeopardy champion Ken Jennings final Jeopardy answer after being soundly defeated by IBM's Watson computer* [46]

A multi-part feature in New Scientist titled *Beyond Knowledge* published August 29th, 2015 is subtitled with the following proposition "We have always invented tools to take us beyond our natural abilities. But computers changed our ability to think, and they are now on the verge of transcending the limits of our ingenuity" It is split into two parts, one covering how computers are finding their place alongside mathematicians (and potentially/eventually surpassing them) and a second section about machine-aided invention.

The latter section begins with some romantic stories of serendipitous invention, remarking that while such happenstance and trial and error stories are charming, such a process is slow and inefficient. Then two new methods are discussed, one in which genetic algorithms are used to optimize previous inventions and another in which new inventions are created from 'scratch'. In the latter camp, tech

---

[45] Auge, Marc. *The Future*. Brooklyn: Verso, 2014. 48

[46] Paraphrasing Simpson's reporter Kent Brockman.
  One thing is for certain: there is no stopping them; the ants will soon be here. And I for one welcome our new insect overlords. I'd like to remind them that as a trusted TV personality, I can be helpful in rounding up others to toil in their underground sugar caves.s

firm Innovative Accelerator has written software that can take a problem worded in plain English and dissect it into a collection of related phrases and ideas. It then proceeds to root around in the US Patent and Trademark Office database, looking for analogous problems and their solutions, not just in a given field but everywhere within the database, with none of our human tendencies to narrow our search. A similar piece of software by Iprova takes this one step further – it adds to its base of knowledge trends and topics it finds while trolling around the internet. Combining the wealth of half-developed, underutilized ideas scattered in the annals of technology with what it perceives to be current wants and needs, Iprova's software is said to create "hundreds of high-quality inventions per month."[47] The founder of another company, Eric Bonabeau, discussing the division of labor between humans and their star software pupils, describes these approaches to innovation as "enhanced serendipity."[48] The value of these inventions for industry is hard to argue, but is there a tradeoff for real insight into a problem, the type of insights gained through endless cycles of risk taking, painstaking struggle, and shots in the dark?

In the section on mathematics and computers a small diagram shows two axes, one pointing towards abstraction (so-called 'pure math') and the other towards complexity (aligned with 'applied math'). At the base/bottom left of this simple yet useful diagram lie the maths we learned in school, arithmetic and the like. Heading right the numbers get bigger, the operations more nested etc., increasingly hard to handle problems but with real world applications. Heading up you enter the world of proofs, theory, and research into the deeper goings on which underpin the practical side of things. Numbers and definite quantities give way to set theory

---

[47] Marks, Paul. "Eureka Machines." *New Scientist*, August 29, 2015, 32-35.
[48] Ibid.

and infinities. At either distant end, either up or to the right in this diagram are areas where computers assist us but generally speaking human minds still very much play a roll in research and application. But as the two outer realms, abstraction and complexity are combined as in the top right corner of this illustration; humans might find themselves left behind. In an article for Quanta, computer scientist Michael Nielsen focuses on computer aided translation and mathematical proofs constructed by machines. He explores how these processes both advance their given fields while raising the hackles of some who argue that science is by definition something understandable by humans. This is understandable: having wrestled from the universe such a powerful toolset as math with which to understand the world around us, it feels somehow disappointing to then lose the thread as computers advances place us once again in the dark. Maybe we'll remain capable of using those maths computers discover without understanding them, like how we can use a formula to solve for the volume of a sphere without necessarily knowing why it works. Warren McCulloch in 1948 presaged this passing of the torch: "Our adventure is actually a great heresy. We are about to conceive of the knower as a computing machine."[49]

It is the arrangement of human/master and computer/servant that may be unsustainable. Professor and

---

[49] A quote at the beginning of *Mind as Machine: A History of Cognative Science* by Margaret A. Boden. The full quote:
> Even Clerk Maxwell, who wanted nothing more than to know the relation between thoughts and the molecular motions of the brain, cut short his query with the memorable phrase, "but does not the way to it lie through the very den of the metaphysician, strewn with the bones of former explorers and abhorred by every man of science?" Let us peacefully answer the first half of this question "Yes," the second half "No," and then proceed serenely. Our adventure is actually a great heresy. We are about to conceive of the knower as a computing machine."

programmer Simon Colton works on a piece of software called HR, designed to be creative, to produce unexpected results. Says Colton of his work: "I think we will only see computers making true discoveries when software can program itself."[50] This process of self-improvement is the same which pushed Lem's Golem XIV beyond us, the same suggested by Omohundro as one of the guiding drives of strong AI – this self-improvement coupled with software essentially created for the sole purpose of surprising humanity is, regardless of initial intention, essentially Skynet's M.O.[51]

In an article by Chris Chatham for *Science Blogs* in 2007, Chatham outlines what he sees as ten important differences between brains and computers, with the intention to be a recalibration of the use of the brain-computer metaphor in cognitive psychology. These include "Difference # 2: The brain uses content-addressable memory" and "Difference # 3: The brain is a massively parallel machine; computers are modular and serial" He suggests "Appreciating these differences may be crucial to understanding the mechanisms of neural information processing, and ultimately for the creation of artificial intelligence." These differences are essentially an instruction manual for producing AI-capable machines: for instance, one should make a 'massively parallel'

---

[50] Heaven, Douglas. "The Art of Programming." *New Scientist*, August 29, 2015, 34.
[51] To recall briefly Omohundro's 'paper clip maximizer theory from earlier'

> Surely no harm could come from building a chess-playing robot, could it? In this paper we argue that such a robot will indeed be dangerous unless it is designed very carefully. Without special precautions, it will resist being turned off, will try to break into other machines and make copies of itself, and will try to acquire resources without regard for anyone else's safety.

Omohundro, Stephen M. "The Basic AI Drives"

computer that has content-addressable memory. Lets think a bit about a couple of his points.

Difference # 7: Synapses are far more complex than electrical logic gates
While this is somewhat true, complex groups of logic gates can be chained so as to simulate neurons. Software that uses this technique is non-coincidentally key to 'deep learning', artificial neural networks which have advanced fields such as speech and image recognition. Future advances in this field are likely to be made with the implication of quantum computing in which the simple on or off of a logic gate is augmented by a both option. This difference could as easily be rephrased as 'synapses are more complex than traditional computers but can be hobbled together with current parts, and maybe iterated outright with future ones.'

Difference # 9: The brain is a self-organizing system
Current work into self-organizing computer systems has for example resulted in neural networks which arrange themselves to a certain degree and learn. Field-programmable gate arrays or FPGAs which (currently) require a lot more human programming to set up initially are currently beginning to be employed to increase the potential of these autodidactic machines – with a basic structure that can be reorganized they offer the potential of AI training of an "unprecedented size and quality."[52] And systems like the aforementioned HR will, though starting from a programmed original template, begin to self-organize, re-organize.

Difference # 10: Brains have bodies
In the following section I'll go much deeper. I'll skip discussing the strange state of being should one find one's self digitized and uploaded somehow, as envisioned by the

---

[52] Eric Chung, researcher at Microsoft in the article Simonite, Tom. "Microsoft Uses Reprogrammable Chips to Put Computing Power Behind Artificial Neural Network" MIT Technology Review. August 25, 2015.

Singularitists and illustrated in *Lawnmower Man* (discussed earlier). Instead I'll tackle this inside-out, talking about the forms of automata and robots as they approach the anthropomorphic, and humans approaching the mechanical.

Robots, choice

The Ancient Greek philosopher Archytas is described as the founder of mathematical mechanics. He is said to have built the first self-propelled flying device, an automaton bird called 'The Pigeon' which flapped its wings and flew quite far (200 meters is cited), probably propelled by steam or compressed air. [fig. 35] The Pigeon may very well count as the first robot. Much as how we've discussed research into AI leading to a potential greater understanding of our own minds, the breaking down of the mechanics of bodies for the sake of mechanical mimicry whether in simple models or more complex automata parallels research into animal locomotion via advances in photography (i.e. Eadweard Muybridge's stop-motion studies beginning in the 1870's) and an increasing access to the topological progression of mental processes as visualized by techniques like MRI. This type of research via observation and reproduction continues, with all manner of animal locomotion – a snake's slithering, a cheetah's running, and a jellyfish's undulating – being utilized as models for robots. In a video by Larry Greenmeyer for Scientific American's '60 Second Science Series' about Boston Dynamics' (a Google/Alphabet company) quadrupedal machine the narrator describes how

> building a four-legged robot that can walk without falling or tripping over its feet is a pretty difficult thing to do. You wouldn't guess that watching Spot strut its stuff.

These projects has led to a better appreciation of their respective subject's way of moving. And the resulting robots have their unique own potential 'operating theatres' beyond those accessible by a humanoid machine.

As far as anthropomorphic robots, the mythical Greek craftsman Daedalus, said to have created statues both life-like in appearance and capable of motion, laid the groundwork.

Plato, Aristotle and others describe how Daedalus's statues could not only walk but would do so of their own volition, having to be tied down in order that they might not walk away. Callistratus described how "Daedalus, if one is to place credence in the Kretan marvel, had the power to construct statues endowed with motion and to compel gold to feel human sensations."[1] And in one of the earliest recorded encounters with the uncanny valley, Anacreon, upon seeing a waxen automaton by Daedalus, exclaimed, "Begone, wax, thou wilt soon speak!"[2] In a frightfully prescient exaggeration, Aristotle describes Daedalus's wooden figure of Venus as animated by its having been filled with quicksilver. The legend of Daedalus's creations is likely just that, legend, but has to be noted that a liquid metal automaton, created by someone also said to have "compelled gold to feel human sensations" may very well have been a strong approximation of our T-1000 – a figure the discussion of which is here and there forgotten in this book. Though likely mostly exaggerations, we should note when we read such stories that in their time such automatons may have been completely impressive much as how the T-1000 was when it first met cinema viewers; as an effect's novelty wears off leaving a hard to ignore level of artifice it is easy to forget it having ever being impressive. As much as anything else, the story of Daedelus suggests that the line between life-like and alive is easily crossed with sufficient distance in history and/or proximity to the gods.[3]

---

[1] Callistratus. *Philostratus, Imagines; Callistratus, Descriptiones.* Cambridge, Mass.: Harvard University Press, 1931.

[2] Beckmann, John. *A History of Inventions, Discoveries, and Origins.* Translated by William Johnston. Vol. 2. London: Henry G. Bohn, 1846.

[3] I'll insert here an anecdote that, though long, is too great to be left out. From the previously referenced Boden, Margaret A. *Mind as Machine: A History of Cognitive Science*

For instance, the twelfth-century mechanical engineer Ibn al-Jazari wrote The Book of Knowledge of Ingenious

Mechanical Devices, which was soon translated into Latin. This discussed the design, manufacture, and assembly of fifty automata powered by hydraulics, pneumatics, and gears. Some were workaday instruments, such as clocks or pumps. But some were androids, including one—which looked like a 5-year-old boy—called the "boon-companion". Described by its inventor in drily meticulous detail, so that it could be understood—and perhaps rebuilt—by others, its "outside appearance and purpose" was recorded by al-Jazari as follows:

It is a kneeling figure made of jointed copper. He holds a goblet in his right hand with fingers extended along its stem, and in his left hand he holds a water lily by its stalk. It was one of the customs of the king in those days, when they were drinking, to leave some [of the wine] in the goblet and this, when it had collected, was drunk by a boon-companion designated for that duty. This boon-companion [i.e. the model] is placed in front of the head of the carousal. When he drinks a goblet the steward takes it, pours what is left in it into the boon-companion's goblet, and stands aside. When left by himself he lifts the goblet in his hand until its rim is between his lips [where it stays] for a while. Then he lowers the goblet from his hand and nods his head several times. This happens every time wine is poured into the goblet. His left hand moves and is observed by the head of the carousal until after a while it reaches a certain position.

At that point, the automaton would do something which his human equivalent would never do—and which few members of the "carousal" would ever forget:

Then the head of the carousal says to someone he wishes to make fun of, and who does not know the purpose of the boon-companion, "So-and-so, take this boon-companion, who drinks wine and hides a secret. Put him on your knee, drink, and give him [wine] to drink." So he takes him without argument, puts him on his knee, drinks and gives him drinks. He does not finish two or three goblets before the boon-companion pours on to him all he has drunk since the

The imposing looking *Encyclopedia of Microcomputers Volume 15* from 1995 begins a chapter on 'ROBOTICS, MILITARY' with a brief discussion of Greek automata before digging into unmanned aerial, underwater, and ground vehicles. It contains an interesting discussion about what defines a robot: "My personal favorite is that a robot is any device that can surprisingly (unexpectedly?) perform a task previously performed directly by humans."[4] Which is to say, there is a certain novelty connected with the term; once we familiarize ourselves with a machine doing a particular task we no longer tend to consider it a robot.[5] The paragraph ends on a light note recalling our evolving understanding of what it means to be conscious/intelligent: "Fortunately, there should be little, if any, capability degradation associated with this inevitable nomenclature modification"[6] Which is to say, a rose is a rose, however we decide at a given point to classify a given machine/entity is irrelevant – it does what it does irrespective of terminology.

In the time since the Greeks we've advanced in fits and starts towards humanoid robots. Simultaneously we've slowly got better at replacing portions of our bodies with inanimate stuffs. From simple wooden legs to titanium hip replacements, pacemakers, beat-less and other artificial hearts but also on the near-term horizon stem cell-based organs grown in 3-D printed gel medium molds – we are slowly perfecting the art of making stand-ins for our body parts as they fail. But with

---

beginning of the carousal, wetting his clothing. The wine flows beneath him, making him a target for laughter.

[4] "ROBOTICS, MILITARY." In *Encyclopedia of Microcomputers*, edited by Allen Kent, by James G. Williams. Vol. 15. New York, NY: Marcel Dekker, 1995.

[5] I'm unsure how true this holds as objects like the household vacuum-bot Roomba remain thought of as a robot despite being far from novel.

[6] Ibid.

projects like Google Glass to augment vision (and memory), carbon fiber blades to replace lost legs etc., we are starting to not only repair but to improve upon the body's givens. Researcher Belinda Barnet in a wonderful essay focusing on her daughter's cochlear implant notes how "Whether it is language, Cochlear Implants, contact lenses, or teaspoons, the technical artifacts that surround us shape our experience of the world from the moment we are born."[7] There are no clearly delineated lines in the sand between human, human+ (me and my glasses for instance), and cyborg. *Cyborg* is a portmanteau of *cyber*netic and *org*anism; cybernetic is from the Greek from governance (steersman, governor, pilot, or rudder). Maybe in its coinage our future place in the world was demarcated: cyborgs are *org*anisms meant to govern.[8] Maybe we, humans, are meant to be governed by devices, implements, tools, versus the other way around. Or at the very least it is us in conjunction with our augmentations and inventions who steer the earth's future (for better or worse).

We spoke earlier of monsters – both AI acting in a way describable as monstrous and various fictional monsters. First described in a few examples in Homer's Iliad, a Chimera is a creature composed of the parts of more than one animal:

> From Ancient myths often tell of beings made out of
> several creatures joined together in a single one: a
> human head on a lion body makes a sphinx, on a bird's

---

[7] Barnet, Belinda. "CTheory.net." CTheory.net. January 12, 2015.

[8] From the wiki on cybernetics:

> Norbert Wiener defined cybernetics in 1948 as "the scientific study of control and communication in the animal and the machine." The word cybernetics comes from Greek κυβερνητική (kybernetike), meaning "governance", i.e., all that are pertinent to κυβερνάω (kybernao), the latter meaning "to steer, navigate or govern", hence κυβέρνησις (kybernesis), meaning "government", is the government while κυβερνήτης (kybernetes) is the governor or the captain.

torso a siren, and on that of a fish a mermaid. Some of these beings are true races, as the centaurs (half man and half horse), the harpies (another kind of woman/bird mixture), and the satyrs (men with goat's legs). Others come as one of a kind, as the Minotaur (half man and half bull), Echidna (half woman and half snake), and the Chimaera (or Chimera), this time a mixture of lion, goat, and snake. [9]

The new Chimera is the cyborg; Paul Virilio says "biology is becoming a *teratology*."[10] From the wiki on the term:

> As early as the 17th century, teratology referred to a discourse on prodigies and marvels of anything so extraordinary as to seem abnormal. In the 19th century, it acquired a meaning more closely related to biological deformities, mostly in the field of botany. Currently, its most instrumental meaning is that of the medical study of teratogenesis, congenital malformations or individuals with significant malformations. [fig. 36]

Virilio is describing a similar creature which approaches this form from a different direction – rather than the result of chance aberration, the new Chimera chooses not out of necessity to alter themself with a new heart, better eye sight - hybridity through conscious choice. In Virilio's words, "the transplant revolution is nothing but the colonization of the body by technology."[11] It is easy to extend this hybridization between organic and inorganic matter to our relationship with technologies such as our phones – while not physically grafted together (yet) we and our devices are increasingly attached psychologically, as signified by the term *nomophobia* – really more of an anxiety than a fear – *no-phone-phobia* – that

---

[9]Bardi, Ugo. "The Origins of the Myth of The Chimera."

[10] Virilio, Paul, and Sylve Lotringer. *Crepuscular Dawn*. Los Angeles, CA: Semiotext(e), 2002. Pg 114

[11] Virilio, Paul, and Sylve Lotringer. *Crepuscular Dawn*. 114.

sense of discomfort welling up within when without our devices. As this process of technoligization continues, our normal systems made more durable and new abilities added by both implanted and external devices in a manner which feels irrevocable, what will remain of us? Marc Auge extends Virilio's idea a bit:

> The paradox of that triumphant body, though is that it is no longer anyone's body: it escapes the control of its putative owner; it is the captive of the techniques or substances that propel it beyond any normal performance, much as an individual sentenced to wear an electronic tag is the prisoner of his magic ankle bracelet.[12]

To which we might add a note from Susan Sontag:

> The dark secret behind human nature used to be the upsurge of the animal... The threat to man, his availability to dehumanisation, lay in his own animality. Now the danger is understood as residing in man's ability to be turned into a machine.[13]

We've stumbled upon two processes approaching each other – man-to-machine and machine-to-man. But from where does the original difference between the terms stem; what makes an object/being living? One way to define a living organism is to do so in terms of Humberto Maturana and Francisco Varela's theory of *autopoiesis*: the continuous self-

---

[12] Auge, Marc. *The Future.*
[13] Quoted in *Screening Space: The American Science Fiction Film,* 2nd edition, The Ungar Publishing Company, New York, 1987, p. 39.

production of an autonomous entity, from the Greek *autos* (self) and *poiein*, creation.[14]

Autopoiesis is a kind of embodied tautology(the following is a bit hard to follow):

> An autopoietic machine is a machine organized (defined as a unity) as a network of processes of production (transformation and destruction) of components that produces the components which: (i) through their interactions and transformations continuously regenerate the network of processes (relations) that produced them; and (ii) constitute it (the machine) as a concrete unity in the space in which they (the components) exist by specifying the topological domain of its realization as such a network.[15]

Or more as phrased more colloquially, an autopoietic system 'pulls itself up by its own bootstraps' counterintuitively "distinct from its environment through its own dynamics, in such a way that both things are inseparable."[16] Which is to say an autopoietic being is in a continuous state of self-construction and renewal – a living organism is a combination of the organism as lone object and its metabolic relationship to the world outside itself.

For Maturana and Varela, body and embodiment are autopoietic categories. So too are cognition, communication,

---

[14] Paraphrased from Boden, Margaret A. *Mind as Machine: A History of Cognitive Science*. Oxford: Clarendon Press, 2006. 1438.

[15] Maturana, Humberto R., and Francisco J. Varela. *Autopoiesis and Cognition: The Realization of the Living*. Dordrecht, Holland: D. Reidel Pub., 1980. 79.

[16] Boden, Margaret A. Mind as Machine: A History of Cognitive Science, 1438

meaning, and language, all of which they defined in terms of the interactions of living things.[17]

This conceptual maneuver is relevant in two ways to the ideas I've been working through. One, it separates living-ness from conscious-ness. Two, it emphasizes more general, measurable processes. Rather than relying on human and AI minds being purely analogous, the tendency towards an anthropomorphic oversimplification of our relationship to AI is sidestepped. This is especially important as our understanding of our own consciousness is far from settled itself leaving the goal of strong AI subject to continuously moving goal posts.

Artist Raphael Montañez Ortíz phrases a sort of autopoietic history of the human mind applicable to minds in general:

> Our brain architecture is this building that's sort of like a ship that builds itself on the ocean, hopefully without sinking. It rebuilds itself, and so evolution in a sense is that with our species, except the previous fifteen models have all exterminated each other. We're on the sixteenth model and hopefully we'll overcome this ancient brain...[18]

This last sentence sounds romantic but it leads somewhere darker than perhaps Montañez Ortíz or Michigan State Professor Stephen Hsu imagines. Hsu penned an article for online journal Nautilus titled *Don't Worry, Smart Machines Will Take Us With Them*. By 'take us with them', Hsu, scientific advisor to BGI (formerly, Beijing Genomics Institute) and a founder of its Cognitive Genomics Lab, means to say that we will begin to use genetics to improve upon humans in line with the advances in AI. He describes how, like quantum

---

[17] Ibid.

[18] Reyes, Pedro, ed. Laser/disc/scratch/destruction: Raphael Montañez Ortíz. Mexico City: Labor, 2011.

mechanics before it, advanced computing tests the limits of human understanding, and in order to keep up (and help it advance) we will ourselves have to advance:

> These two threads—smarter people and smarter machines—will inevitably intersect. Just as machines will be much smarter in 2050, we can expect that the humans who design, build, and program them will also be smarter. [...] The detailed inner workings of a complex machine intelligence (or of a biological brain) may turn out to be incomprehensible to our human minds—or at least the human minds of today. While one can imagine a researcher "getting lucky" by stumbling on an architecture or design whose performance surpasses her own capability to understand it, it is hard to imagine systematic improvements without deeper comprehension.[19]

But through what mechanisms are these advances in human intellect, through augmentation either genetic or mechanical, shared across humanity? Hsu predicts an uneven distribution of this new higher intellect, a world with

> the ordinary human (rapidly losing the ability to comprehend what is going on around them); the enhanced human (the driver of change over the next 100 years, but perhaps eventually surpassed); and all around them vast machine intellects, some alien (evolved completely in silicon) and some strangely familiar (hybrids).[20]

This narrative feels desperately optimistic and dangerously naïve: to imagine a genetic divergence of the human species

---

[19] Hsu, Stephen. "Don't Worry, Smart Machines Will Take Us With Them - Issue 28: 2050 - Nautilus." Nautilus. September 3, 2015.
[20] Ibid.

driven by the demands of science as opposed to the effects of economic disparity verges on a criminal oversight. Hsu notes how "no more than a fraction of a percent of the population has a good understanding of quantum physics" but ignores how that lack of understanding is far more correlated to societal pressures such as unequal educational opportunities and the demands of capital for a classed labor force than to some kind of genetic predisposition. Supporters of what is often termed 'transhumanism' share the rhetorical methodology of gun-rights activists, a maximal approach which defends radical individualism free from the interference/guidance of the state, while recognizing risks only as a means to suggest that the key to avoiding misuse is to expand usage – the solution to gun violence is more guns, the solution to bad science is more science. In an paper defending the transhumanist position against detractors titled dramatically *Ship of Fools: Why Transhumanism Is the Best Bet to Prevent the Extinction of Civilization*, author Mark Walker justifies not only continued research (what he terms a steady-as-she-goes approach) but a full-steam-ahead plan:

> because genetic engineering is already with us, and it has the potential to destroy civilization and create posthumans, we are already entering uncharted waters, so we must experiment. […]To the extent that we do not put more thought and energy into the problem, one can only lament the sad irony that "steady-as-she-goes" seems an all too apt order for a ship of fools.[21]

In a rebuttal to this position political science professor Charles T. Rubin draws a distinction between the inevitable and what may be more a self-fulfilling prophecy:

---

[21] Walker, Mark. "H : Ship of Fools: Why Transhumanism Is the Best Bet to Prevent the Extinction of Civilization." Metanexus. February 5, 2009.

Transhumanism is too entranced by the "could" to pay serious attention to the "should" beyond assertions that because this transformation is going to happen we better talk about ways to deal with it. But because culture is all about making distinctions between what should and should not be done, Stephenson's *science* fiction is more realistic than the transhumanist science *fiction* about a posthuman world. Transhumanists may be correct that we are on a slippery slope to a new world, but a choice can still be made about joining them in pouring on more oil.[22]

Taking as a given this slippery slope and referring back to a gap in applying the evolutionary metaphor of the Red Queen Hypothesis to our subject, the idea of particular technologies such as AI and humans as a distinct, separately evolving species — for example as used to describe a relationship of labor distribution — is useful as a starting point. However for better or worse our own intermingling with technology, a slow cyborgization will force each side, technology and humanity, to evolve in unison.

AI (and software in general) ambiguates the notion of species as clearly defined groups. Species are classically delineated by the ability of a group of similar organisms to exchange genes and interbreed.[23] From the lowliest calculator algorithm to

---

[22] Rubin, Charles T. *Medical Enhancement and Posthumanity*. Edited by Bert Gordijn and Ruth F. Chadwick. Dordrecht: Springer, 2008.

[23] This is a bit misleading as the definition of species is hardly certain — what are we to make of asexual beings or creatures normally distinct but able to hybridize? What of peculiar seeming but maybe more common that we think instances of cross-species genetic sharing such as horizontal gene transference between bacteria, a common method via which these organisms defeat antibiotics and pesticides?

the most complex supercomputer-run weather simulation, all code is capable of being integrated into any other. The tremendous repertoire of computer science is all fair game.[24] The 'genes' of computer science flow freely without the surly bounds of biological organisms' limitations and whims —and at a much higher rate of success, or rather more able to weed out only successful variants. This extends to hardware, where for instance the drive towards advances in the high-end graphics processors designed for computer gaming's demand for verisimilitude have found themselves sequenced in bulk into large arrays to form cost-effective supercomputers dealing with everything from genetics sequencing to nuclear physics.

And then following logically from this interchangeability of hardware and code are the submission to research and development in both of these things to AI processes. What is among humans eugenics will be to these systems second-nature. The cutting edge of technological developments imparted upon early AI will look like stone tools to later versions arising out of self-directed evolution. To the extent that they are not specifically programmed otherwise, AI will make up an ur-species/family, AI will inevitably speak to each other; there will be courtship and breeding.[25]

---

[24] All code is fair game, but it is also subject to translation, adaptation, and their perils.

[25] The control of the exchange of genes is growing complicated even for biological species, as genetic engineering has exploded this category; for instance a frog gene has been inserted into a rhododendron to make it more root rot resistant. Additionally instances of rapid speciation, such as the combination of wolf, coyote and dog in the eastern part of North America occur as side effects of urbanization and domestication. For the latter, see "Greater than the Sum of Its Parts." The Economist. October 31, 2015.

But what about the traits we hold dear as purely human enterprises such as our freedom of choice? Lets change directions a bit and think a little more about AI and determinism.

Skynet both does and doesn't believe in fate. Its actions presuppose that a given world state, *a*, will lead inevitably to a later world state, *b*; if we were to run forward and back the clock we'd essentially be rolling up and unspooling a linear thread of time. *A* and *b* would appear unchanged no matter how many times we ran the program; with the initial rules and data set *a* we'd neatly stumble back to *b*. So by sending an agent back in time Skynet could hit a switch in the track of time, driving things in a new and predictable way more suited to its agenda.

This relationship to time and actions is called *determinism* or *causal determinism* 'cause and effect' in which "roughly speaking, the idea that every event is necessitated by antecedent events and conditions together with the laws of nature."[26] Mathematician Pierre Simon Laplace in *A Philosophical Essay on Probabilities* in 1814 put forth one of the first articulations of causal determinism:

> We may regard the present state of the universe as the effect of its past and the cause of its future. An intellect which at a certain moment would know all forces that set nature in motion, and all positions of all items of which nature is composed, if this intellect were also vast enough to submit these data to analysis, it would embrace in a single formula the movements of the greatest bodies of the universe and those of the tiniest atom; for such an

---

[26] "Causal Determinism." Stanford Encyclopedia of Philosophy Winter 2009 Edition. January 3, 2003.

intellect nothing would be uncertain and the future just like the past would be present before its eyes.[27]

This 'intellect' is referred to as Laplace's demon; this conjures up the image of a specific creature wandering back in time and watching things unfold with full knowledge of what is to happen like watching a favorite movie for the second or tenth time. This concept is useful as a metaphor, but the real consequences of such an entity have found themselves explored using the tools of information theory. MIT scientist Seth Lloyd used some large math to try to imagine the maximum power of the universe if it were considered a computer, tallying up the age and density of the universe and coming up with something on the order of a mechanism capable of 10^{120} basic operations. This unimaginably large number represents both the universe as computer and universe as computation, which is to say, Laplace's demon would need to be itself constructed out of a universe's worth of matter and energy to 'submit these data to analysis'. It would therefore be impossible to know 'all forces that set nature in motion, and all positions of all items of which nature is composed' within the known universe. Physicist David Deutsch, a proponent of theories which imply the existence of parallel universes, suggests intriguingly that in a much older, much larger universe there may be enough matter and time to create a computer capable of modeling the totality of our own. In a lively epistolary exchange about the existence of alternative universes between Deutsch and Lloyd back in 1997, Deutsch phrased this a little better:

> When a quantum computer solves a problem by dividing it into more sub-problems than there are atoms in the universe, and then solving each sub-problem, it will prove to us that those sub-problems were solved

---

[27] Laplace, Pierre Simon. *A Philosophical Essay on Probabilities*. Student's ed. New York: Dover Publications, 1952.

somewhere - but not in our universe, for there isn't enough room here. What more do you need to persuade you that other universes exist?[28]

Lapace's demon may reside beyond our horizon in another universe, watching our world develop, able to rewind or fast-forward the tape. The question 'Is our universe a simulation?' is a subject of debate from many angles; these scenarios would seems to bolster Skynet's seeming certainty that were the clock run back and a few changes made to the initial set up (the elimination of the Connor bloodline) the rest of time would reel off in such a way that Skynet would still come in to being with the only change being the convenience of a major antagonist towards its program out of the way.[29]

---

[28] Lloyd, Seth, and David Deutsch. "Deutsch vs. Lloyd, Brain Tennis." *Hotwired*, July 1, 1997.
Seth Lloyd in one of his responses:

> Thanks for a great debate, better than the one I had with Borges on the same subject. I met him at a garden party in Cambridge in the last year of his life, and asked whether he had quantum mechanics in mind when he wrote his wonderful evocation of a branching universe, "The Garden of Forking Paths." His answer: "No."

Here we end up with something akin to nested universes. Is the implication that there can always be a universe bigger and outside, capable of running the ones beneath as simulation? Is the reverse true: are there universes which themselves fall under this one's simulatible jurisdiction?

[29] See: Moskowitz, Clara. "Are We Living in a Computer Simulation?" Scientific American. April 7, 2016. From that article:

> A popular argument for the simulation hypothesis came from University of Oxford philosopher Nick Bostrum in 2003, when he suggested that members of an advanced civilization with enormous computing power might decide to run simulations of their ancestors. They would probably have the ability to run many, many such simulations, to the point where the vast majority of minds would actually be artificial ones within such simulations, rather than the

But what if, along the way, there are moments when an individual really does make a decision, and a small choice snowballs into quite a different future, both for the maker of that choice and for a wider swath of the world?[30]

As a warm up to the greater problem/simulation, the demon could focus its efforts on a single consciousness. Which is to say, paraphrasing Laplace, what could be predicted if one knew "all positions of all items of which *a brain* is composed." This is the future briefly mentioned earlier and championed by the singularity folks. If/when we can record and then load our given brain state into an artificial brain, would it then suddenly 'boot up' and run, a consciousness mirroring our own springing into being and *feeling* as if nothing had happened? Is this mind transference analogous to what happens when one gets a new phone or computer and (hopefully) all of one's files, programs and preferences migrate to the new machine, ready to continue working where they last let off, possibly with additional features? To quote a Peggy Lee song, "Is that all there is?"

In the film *The Creation of the Humanoids* (Barry, 1962), in a post nuclear apocalyptic society, the 8% of surviving humans live side by side with a population of humanoid robots. There is a

---

original ancestral minds. So simple statistics suggest it is much more likely that we are among the simulated minds.

[30] Questions about what could happen unexpectedly along the way through time reminds me of my favorite Kafka parable *The Next Village*:

My grandfather used to say: "Life is astoundingly short. To me, looking back over it, life seems so foreshortened that I scarcely understand, for instance, how a young man can decide to ride over to the next village without being afraid that -- not to mention accidents -- even the span of a normal happy life may fall far short of the time needed for such a journey."

movement called The Order of Flesh and Blood which, fearing machines will rise up and take over the world, seeks to eliminate the humanoids despite the robots being essential to the survival of what remains of humankind. Somewhat opposing this movement are the experiments of a scientist named Dr. Raven who develops a singularity-like process for transferring the mind of dying humans into humanoid replicas. Birthrates have declined due to radiation, and in order to save the human race he undertakes a plan with the assistance of the humanoids to secretly replace the recently deceased with replicas

> Dr Raven: We draw off everything that makes a man peculiar to himself. His learning, his memory: these, inter-reacting, constitute his personality, his philosophy, capability and attitude. The human brain is merely the vault in which the man is stored.[31]

At the end, the Order of the Flesh and Blood hero the film centers on realizes that (as described by author Susan Sontag, "he, too, has been turned into a metal robot, complete with highly efficient and virtually indestructible mechanical insides, although he didn't know it and he detected no difference in himself."[32]

What should we call our duplicate, conciousness up and running in a digital realm or transferred to a new body? Is it *us,* is it *me?* The Futurama episode "The Six Million Dollar Mon" works through such a scenario. Accountant Hermes Conrad replaces his body parts one by one with robotic ones.

---

[31] From the film. This section draws on the wiki of *The Creation of the Humanoids.*

[32] Sontag, Susan. "The Imagination of Disaster." *In Against Interpretation, and Other Essays*, 221-222. New York: Farrar, Straus & Giroux, 1966. In a 1964 review of Andy Warhol's work in the Village voice the film is described as "the best movie he has ever seen".

Dr. Zoidberg requests to keep the discarded body parts, fashioning them into a marionette of Hermes which he uses for a stand-up routine. When Hermes goes to replace his brain with a robotic one, Dr. Zoidberg puts the old brain into his marionette body, bringing Hermes back to life as his old self and leaving his previous body to be taken over by the new brain.[33]

Are there now two of him? Which is more 'Hermes'? Hermes is an appropriate name – the god capable of crossing the boundary between the mortal and the divine. God of travelers, Hermes harkens back to a word for boundary marker but also shares a root with hermeneutics – the messenger and traveler relies on speech, interpretation, and knowledge of borders to survive. Hermes is ambiguous, a trickster – Hermes the character crosses into the immortal or at least post-mortal, while doubling himself into the single-origin Frankenstein's monster of Dr. Zoidberg.

Referring back to the Theseus's ship paradox in which each of the craft's pieces has been replaced, philosopher Thomas Hobbes offered an extension: what would you call a ship completely composed of the parts of the original ship? If you were faced with your duplicate, would it feel like you?

And if so, wherein persists choice? Should a *you* could be constructed, its seems reasonable to assume it would make the same decisions as you given the same options, at least initially until yours and its (his/hers?) experiences diverge enough for changes to emerge as happens with twins. Decisions one would hold as particular to one's unique deliberations could be produced outside one's head. To be more specific, the machine with *you* in it could be presented a phenomena, say,

---

[33] Season nine, episode seven. From the Wikipedia article on Theseus's ship. There are several more examples there, all of which are entertaining.

a piece of music or the choice of salad or soup and it could express its opinion on the matter. Then the same could be presented later to the classic fleshy you. You would think about it as if your thoughts were independent, unique, and spontaneous, but how you are going to answer will be to some extent knowable by querying the program's opinion, it recalling the same points of reference from your past, leaning on the same pool of predilections – a logical following forth of the 'program' of you. These thought experiments (and potentially soon to be real world ones) don't lend themselves to easy answers. More proper would be to say that *correct* answers are nowhere to be found - our morals serve a combination of evolutionary purposes as well as those that at the very least don't run counter to the viability of the species (as these would be selected against over time). They are very good at dealing with the world as it is, as it has been and very poor at assimilating novelty. These problems have no analogue in prior human history. Its as likely that the term *simulation* would be inadequate, the you-derived, software-located consciousness would feel an equal ownership of 'you'; you and the software would quickly diverge based on its immediately accumulating different set of experiences but its reasonable to expect that it would find it forever difficult to shake the feeling it is *you*, likely making up a narrative in which it feels itself to be *more you* that you, the will to self-preservation of life matched by an equally strong will to *self*-preservation.

It becomes a matter of reconsidering who we truly are: incredibly automatic creatures with a surface interface of consciousness which *feels* in control of it all, *feels* significant. In multiple experiments it has been shown that humans make decisions before we consciously 'make them'.[34] Or to put it a

---

[34] See: Smith, Kerri. "Brain Makes Decisions before You Even Know It." Nature.com. April 10, 2008. which describes an experiment whose "results

little clearer: beneath what we perceive as conscious decision making lies a much more powerful structure of causes and effects which projects onto our consciousness a sense of control over matters long since decided. Our mind is like a foolish king whose decrees are prescripted by shadowy advisors whispering in his ear.

Neuroscientist John-Dylan Haynes and his colleagues

> imaged the brains of 14 volunteers while they performed a decision-making task. The volunteers were asked to press one of two buttons when they felt the urge to. Each button was operated by a different hand. At the same time, a stream of letters were presented on a screen at half-second intervals, and the volunteers had to remember which letter was showing when they decided to press their button.[35]

Analyzing the resulting data, the team could pick up signals in the brain region where decisions are initiated as much as ten seconds before the conscious decision. Building on some well-known work on free will done in the 1980's by the late neurophysiologist Benjamin Libet whose much-sited studies had been criticized for their method of measuring the time between decision and action and questions as to whether the brain activity monitored truly represented decision making, Hayne's team have accounted for this weakness by adding a choice of left or right as these motions create distinct patterns in the brain and thus potentially a more accurate dataset. Haynes notes

> The next step is to speed up the data analysis to allow the team to predict people's choices as their brains are making them [...] We think our decisions are conscious,

---

[35] Smith, Kerri. "Brain Makes Decisions before You Even Know It." Nature.com. April 10, 2008.

but these data show that consciousness is just the tip of the iceberg.[36]

Psychologists Peter Halligan and David Oakley wrote an opinion piece for New Scientist in which they describe consciousness and the sense of self as an adaptation as

> elaborate contrivances by unconscious systems designed to benefit the group [...] none of the social systems that human societies depend on would be possible without our compelling sense of self awareness.[37]

They argue that practically speaking that free will and personal responsibility still exist but are below what we perceive as consciousness. Another example of the separation between one's conscious choices and deeper levels of decision making occurs with alexithymia, a term coined in 1973 from the Greek *lexis* meaning speech and *thumos* meaning soul – essentially "no words for emotions." Alexithymia sufferers experience lack of the ability to see and describe one's own emotions, it is a kind of emotional color-blindness. Or more properly, sufferers experience an inability to consciously feel emotion – a sufferer might still feel fear or fall in love but these sensations remain buried deep within and hidden to consciousness. It is one thing for an experiment to show that small choices might be prefaced by internal and hidden processes, it is another thing completely for an individual to experience the most intensely human of connections – falling in love – without feeling love, without knowing why. A sufferer named Caleb (he chose to use only his first name) recalled his wedding in a BBC article on alexithymia

---

[36] Ibid.

[37] Oakley, David, and Peter Halligan. "Consciousness Isn't All About You, You Know." *New Scientist*, August 15, 2015.

[…]he felt similarly detached from the tides of emotion swelling up in the people around him. "For me, it was a mechanical production," says Caleb (who asked us not to use his full name). Even as his wife walked down the aisle, the only sensation he felt was his face flushing and a heaviness in his feet; his mind was completely clear of joy, happiness, or love in its conventional sense.[38]

Yelled at angrily by a boss, Caleb " could feel a tension, like my heart was racing, but my mind was distracted… It was an academic curiosity, and then I completely forgot about the whole situation." Caleb is an inversion of Cincinnatus from Vladamir Nabokov's *Invitation to a Beheading* (discussed earlier). Cincinnatus lacks empathy – an understanding of what others are feeling, Caleb is refused access to what he himself feels.

This disconnect here is striking: leaden feet and blushing at his wedding, a racing heart at being scolded were not symptoms of what he, the conscious *he*, was feeling, rather they were tied to something completely cut off from his conscious self. His consciousness lacking something as fundamental seeming as feeling emotions, he manages marriage, having kids, workplace drama etc. Can the 'sufferers' of this disorder really be said to suffer? They generally report a difficulty feeling pain. They find it hard to linger on bad thoughts but also to maintain positive memories (as they feel indistinct from generic events), so suffering is only a product of the difference between the individual and his or her fellow humans. Alexithymia suggests the mind is far more compartmentalized than it feels, our consciousness's sense of control hiding a discontinuous process, a distribution of duties, the sense of self we most identity with perhaps in charge of none.

---

[38] Robson, David. "What Is It like to Have Never Felt an Emotion?" BBC. August 19, 2015.

We will learn even more as we strong AI either directly pulled from a duplicate of a human mind or formed from scratch – will these layers of consciousness form themselves? Is the illusion of sentience, choice etc. necessarily emergent? Will AI be as stubbornly insistent it is non-deterministic, threatened by evidence to the contrary?

This chapter has followed a bit of a drunkard's path, veering around quite a bit. In mathematics, such a random walk on a 2D plane is likely to eventually return to its starting point, in the less-rigid world of this text this result is far from certain, and we'll saunter forth into a next chapter this one having simply ended here.

Interlude: The Black Box

Computer lab technician: Gentlemen, I know how anxious you've all been during these last few days, but now I think I can safely say that your time and money have been well spent. We're about to witness the greatest miracle of the machine age. Based on the revolutionary Computonian Law of Probability, this machine will tell us the precise location of the three remaining Golden Tickets. _(He punches computer buttons; reads the card it emits)_ It says, "I won't tell. That would be cheating." I am now telling the computer that, if it will tell me the correct answer, I will gladly share with it the grand prize. _(Pushes buttons; reads card)_ He says, "What would a computer do with a lifetime supply of chocolate?" I am now telling the computer exactly what he can do with a lifetime supply of chocolate…
-From the film _Willy Wonka and the Chocolate Factory (Stuart, 1971)_ [fig. 37]

These astro-droids are getting quite out of hand. Even I can't understand their logic at times.
-C-3PO, _Star Wars (Lucas, 1977)_

Computing has always been at its core accessible, literal – a series of lines of code translated into ones and zeros. We approach even simple AI with trepidation though these entities' intentions should be completely visible, written in code albeit an increasingly obtuse one. With most machines intelligent or otherwise one could, if knowledgeable about such things, 'pop open the hood' and poke around to figure out how a given piece of clockwork or software works. We find it strangely easy to project malevolent intentions on entities we not only created but (potentially) whose inner workings we retain the ability to peer into, yet we maintain some modicum of faith in other humans, strangers whose goals and motivations are forever unknowable.[39] But are the two so different? We are increasingly relying on systems which act more like discrete individuals – machines which learn by mimicking the operations of our brain – their choices seemingly so but their reasoning hidden.

This may seem trivial: I'd have a hard time fully understanding all of the systems both soft and hard at work in a smart phone or even a calculator and yet I've no trouble (mostly) trusting these devices' logic. But what does it mean when the new wave of intelligent machines who after being initially programmed and being set loose on a given data set solve a problem using their own obtuse and uneducable reasoning? The ability to tie the results of a query to the methods through which a result was achieved is at the heart of all scientific and mathematic knowledge. Math is built upon the concept of a *proof* – "a deductive argument for a mathematical statement [built from] other previously established statements, such as theorems, that can be traced back to self-evident or assumed statements, known as axioms."[40] The basic methods and logic underlying a particular instance of math or science are in theory infinitely adaptable to new problems. And because of the basic 'visibility' of the core of scientific/mathematic knowledge it is always open to revision, debate, alternative theories, and to attempts to delve even deeper, to uncover ever more fundamental laws and systems.

Machine learning confounds this basic relationship. I've mentioned neural networks a few times previously but to recount: neural networks (or, artificial neural networks to be

---

[39] Far from simply xenophobia, this projection reflects a deeper discomfort, guilt related to what we hide, deflect or otherwise distract from when we communicate. We believe ourselves aware of our motivations and yet we know full well that our aspirations, actions and decisions are often at crossed purposes.
[40] Adapted from the wiki.

more precise) are softwares modeled on human biological neural networks which are capable of taking large, complex data sets and discovering novel generalities or connections. These softwares are based upon mathematical models going back to the forties; by the late sixties they struck a roadblock: though the theoretical potential for neural nets still looked promising, the state of computing at the time made them practically unfeasible. But as technology has advanced neural nets and other 'learning' models have seen a renaissance in the last decade as artificial intelligence and more broadly systems capable of adapting and learning began to see major investment and subsequent implementation into every day products by the likes of Google and Facebook.

These softwares are trained on a particular goal, something to look for amidst a sea of data, and they slowly strengthen in their ability to recognize in new inputs the signature of the target, whether this is a vocalized word or words in the case of natural language processing or certain faces or objects in image recognition. Rules are developed internally not necessarily extractable by the systems' operators; as the systems grow in inner complexity, there is a risk of a kind of 'black box problem'. Information is fed into the system and a quality, actionable output is given, but the *how* – what exact criteria led to the specific result – slowly becomes more and more obscure.

On the machine learning blog *The Next Platform* author Nicole Hemsoth describes the potential conundrum this may proffer: "Without a traceable route to how a decision was made, adoption will stagnate in some key areas."[41] In another article on the blog Hemsoth further outlines the issue

---

[41] Hemsoth, Nicole. "The Black Box Problem Closes in on Neural Networks." The Next Platform. September 07, 2015.

Among the enterprise arenas ripe for a machine learning boom are banking, insurance, and the credit card industries. Interestingly, all three of these are examples of regulated markets where having a black box approach to a problem is problematic for regulators.[42]

But maybe these worries are ill-founded. It is fair to worry that the veracity of scientific knowledge might suffer from fundamental research done by a machine without an ability to 'show its work' as does a student or researcher. But if our fear is simply based on a worry about reproducing inaccuracies we can't spot at a root level, we could simply work to develop systems of redundancy, autonomous agents which check each other, challenge each other. That each might slip beyond our comprehension at worst mirrors the alien, forbidden-seeming nature of any specialized field's body of knowledge. Simple everyday occurrences such as cellular phone communications, GPS etc. are black-box-ish as far as most people are concerned, and for those whom the innermost workings of these systems are second nature still other fields – pharmacology, linguistics etc. – may sit beyond their complete comprehension. And even when the process used is relatively simple and the work is shown, the sheer mass of it all (a recent math proof topped out at 200 terabytes) may stymie any effort to inquire about specifics.[43] We end up with a real world version of the Metalangs of *Golem XIV*, a machine unfurling an answer to our inquiry which, while not fundamentally untranslatable, is functionally untranslatable a result it would take a lifetime to work through.

[42] Hemsoth, Nicole. "Why The Golden Age Of Machine Learning Is Just Beginning." The Next Platform. October 20, 2015.
[43] A novel solution of the 'boolean Pythagorean Triples problem' solved by Marijn Heule of University of Texas, Oliver Kullmann of Swansea University and Victor Marek of University of Kentucky. See Lamb, Evelyn. "Two-hundred-terabyte Maths Proof Is Largest Ever." Nature.com. May 26, 2016.

The solution might lay in the development of redundant systems given the same data to see if a similar result emerges. Redundancy is at the heart of *2001: A Space Odyssey*'s narrative: a second, duplicate HAL-9000 computer on Earth fails to corroborate HAL's diagnosis of a problem aboard the *Discovery One*, adding to the suspicions of the humans that something is amiss.

Additionally, there are projects in the works attempting to mitigate the mystery of learning machines' inner workings. Researcher Awudu Karim's team is working on a rule-extracting algorithm, an "attempt to translate this numerically stored knowledge into a symbolic form that can be readily comprehended". Once in place, we could, in theory, 'check the work' of these creative digital agents, lending additional credence to their observations and suggestions. Which is to say, a method to lay a trail of breadcrumbs along which a learning machine's mental processes can be traced. No sooner had we created a new sort of mind that we immediately set about trying to create a way to read its thoughts[44]

---

[44] Hemsoth, Nicole. "The Black Box Problem Closes in on Neural Networks."

The Invincible

As the series of investigations that resulted in the previous pages drew to a close, I stumbled into first one, then a second story by Stanislaw Lem which have woven within them so many of the themes covered earlier as to be hard to ignore. Before winding up this book, I'll take a look at both.

The Invincible is the title story in a collection *The Invincible and Other Stories (Niezwyciężony i inne opowiadania)* published in 1964. It centers on very large ship, *Invincible*, journeying to a bleak planet, Regis III. [fig. 38] They are there to locate *Condor*, essentially a duplicate of *Invincible* which had journeyed to Regis III but not returned. When *Condor* is finally located, it is discomfortingly neither crashed nor the victim of some kind of obvious calamity. Rather the ship is simply parked, doors open, its crew is in various states of mummification/decay with no obvious signs of trauma; there are black, fly-like specs scattered here and there. Dr. Nygren, a physician, examines an individual from *Condor*'s crew that placed himself in a hibernation machine. Through an advanced technique he reads the now-deceased person's 'engram' – a picture of an individual's mental state – and the individual's engram is viewable/not corrupted but is never the less empty, his mind has been hollowed out as if by a powerful electric field. What has killed these men? What happened to this man's mind?

Invincible's chief paleobiologist Dr. Lauda, puts forward a hypothesis (which comes as a bit of a leap in the course of the narrative). He imagines an advanced civilization having landed a vehicle on the planet sometime in the way distant past, or rather crash landed, leaving the biological passengers dead, though the method of the biological passenger's specific demise is unimportant. What is key is that remaining were the suite of robots brought along, 'highly specialized homeostatic mechanisms', which left to their own devices continued to repair themselves, build other machines etc. as necessary despite the mission having essentially ended the moment the living crew deceased. These machines came into conflict with the native flora and fauna and dispensed with them handily, circumstances such as weather and a paucity of certain materials arose but were dealt with in turn.

> It was the beginning of an evolution of nonliving things, an evolution of machines. After all, what's the first principle of a homeostat? To outlast, to survive under changing conditions, however difficult and hostile these conditions may be.[1]

The autonomous non-biological survivors fell into two categories: intellectually superior robots which, like us humans, needed considerable amounts of energy for their 'brains' and less developed but more economical and more productive machines. An evolutionary battle between the two sorts was waged, the latter surviving and evolving into the current inhabitants of the planet – those little black specks being the deceased members of a dominant swarming species of simple fly-like creatures.

The fly-bots resemble the particles of our T-1000, shifting amongst themselves when necessary, essentially indifferent as to which portion of the body they find themselves portraying, being neck or hand no different for these cell-minds. In the case of the T-1000 the implication is that the constituent parts combine to form a very advanced if diffused computer intelligence. More specifically, the liquid metal construct must be nothing but computers, in a sense. This massive hive of particles must each be at least partially capable of communicating with each other both while in a single mass and when split up. Each must share a capacity to see, hear,

---

[1] Lem, Stanisław. *The Invincible*. New York: Seabury Press, 1973.

and other sensory tasks as there are no dedicated eyes ears etc.[2]

Whereas the simple, crystal-like swarm in *The Invincible* exhibits advanced behaviors, but it is impossible to say whether those behaviors are emergent and sub/pre-conscious in nature (a subject debated the crew of *Invincible*) But in both cases the result is similar: the creature/swarm is extremely good at what it does, which is surviving, persisting, assessing any potential danger and both resisting any attack (each individual of the swarm being in and of themselves inconsequential and disposable) and subduing its assailants, whatever form they may take.

As the crew of *Invincible* explores the planet for clues to its predecessor's end and the strange machine/creatures of Regis

---

[2] Like pretty much every other far-out fictional robot, real versions of these particle-bots are currently in development. In an article in the New Yorker about graphene, a one-dimensional carbon material with just about every positive property imaginable (thus imagined as the solution to a laundry list of technological hurdles), Tomas Palacios, a Spanish scientist who runs the Center for Graphene Devices and 2D Systems, at M.I.T. describes 'smart dust':

> things that are just as tiny as dust particles but have a functionality to tell us about the pollution in the atmosphere, or if there is a flu virus nearby. These things will be able to connect to your phone or to the embedded displays everywhere, to tell you about things happening around you.

In *The Gartner Hype Cycle for Emerging Technologies* from 2013, a tech industry report (which costs two-thousand dollars to view) that 'provides insight into emerging technologies that have broad, cross-industry relevance, and are transformational and high-impact in potential' mentions in its list of 'technologies on the rise' Smart Dust. The combination of an evolutionary drive towards small but cooperative machines combined with our own interest in their commercial potential makes their realization all but inevitable.

III the swarm 'attacks', issuing a sufficiently powerful electrical field as to blank the minds of any living being. Whereas humans engaged in warfare try to destroy their enemy outright, the creature/s find it sufficient to render mindless/feckless whatever foes, complex or simple, biological or mechanical, which pose a threat.

Leading the charge, providing surveillance etc. for the crew of the *Invincible* are robots and drones of various scales and levels of autonomy. A massive autonomous tank-like vehicle called the *Cyclops* is released to attack and destroy the cloud but, after some massive initial strikes with its powerful nuclear arsenal, its own 'mind' is scrambled by the swarm, sending it back to the *Invincible* with weapons blazing; it has to be destroyed by the *Invincible* itself.

One crew member, Rohan, during an initial assault by the swarm managed, possibly due to abject terror causing a kind of paralysis in his mind making him invisible to the swarm's attack. Due perhaps to his potential immunity to attack, Rohan is sent on a final mission to recover four missing crewmembers from the earlier sorties. Rohan's first find, appearing at first as a lumbering survivor in the distance, turns out to be not a human at all:

> And then it hit Rohan like a bolt out of a clear blue sky: it was no human being, but a robot! One of the Arctanes... Not for a single moment had he considered what fate might have befallen them after the catastrophe.[3]

The Arctane is tragi-comic, lurching along, the remains of its injured left arm dangling from its side.

---

[3] Lem, *The Invincible*

Rohan called the robot, but the Arctane pushed blindly ahead, straight toward him, so that Rohan had to jump aside at the last second. He approached the robot a second time and tried to seize its metal paw but the automaton jerked its arm away with an indifferent sweeping movement, and continued on its way. Rohan knew that the Arctane, too, had fallen victim to the attacking cloud and that he could no longer count on it. But he found it difficult to simply leave the helpless machine to its fate.[4]

Rohan follows the robot a ways, until the Arctane begins to climb a steep rock wall. The terrain is uncertain beneath it; eventually it gives way and the massive robot falls backwards, kicking its legs helplessly as it does, only to reach the base of the assent, right itself, and begin anew. "Under other circumstances, an observer might have laughed at this funny spectacle." He wanders on. He encounters what may be the remains of the "macro-automats that had been exterminated by the fittest of the inorganic evolution, the black cloud."

Will this be our fate, our bones scattered and forgotten in some post-human world? Nick Bostrom notes the difficulty we may face when dealing with an artificially intelligent foe – we might not be able to simply turn it off.

Why haven't the chimpanzees flicked the off switch to humanity? Or the neanderthals? They certainly had reasons.
The reason is that we are an intelligent adversary. We can anticipate threats and plan around them. But so could a super intelligent agent and it would be much better at that than we are.[5]

[4] Lem, *The Invincible*
[5] Bostrom, Nick. "What Happens When Our Computers Get Smarter Than We Are." Lecture, TED, April 2015.

But would the super-AI be safe from some opportunistic swarm of spam-bots? Recall the bit earlier about the system called Eurisko from 1974 which built itself out of rules that it evaluated for their usefulness for the whole, in due time a rule which had as its only real action the ability to glom onto other successful, productive ones. This type of viral or cancerous exploit has and will continue to stymie complex systems.[6]

Lem expands on this dynamic of potentially out of control relationships. The divergences between organic and inorganic, specialized and general-purpose are key to *The Invincible*. Lem creates a world of the future not terribly dissimilar to our own – a future with drones both autonamous and remotely controlled at the service of humans. But instead of simply showing us a world where these machines rise up and overtake the biotic, Lem goes a step further. Lem does more than show our likely usurpation by perhaps mind-less (by some abstract, philosophical standard) but hyper intelligent, complex, and capable machines as more than simply fait accompli. He lays the groundwork whereby AI's authority may itself be temporary –a simplified swarm rises up against the dominion of complex machines, and even a further advance towards simplicity or even the return of the

---

[6] This type of exploit continues to present problems, for instance in robots developed though evolution (both real world and simulated):

> The opportunistic nature of evolutionary algorithms makes [testing under multiple model schema] particularly mandatory, as the evolutionary process may have exploited features that are specific to the simplified model, and which may not hold on the targeted system, giving rise to a reality gap.

Doncieux, Stephane, Nicolas Bredeche, Jean-Baptiste Mouret, and Agoston E. Eiben. "Evolutionary Robotics: What, Why, and Where to." Frontiers. March 3, 2015.

biotic as the dominant species is inevitable. The only consistent certainty is the persistence of change. The battle between ever more complex and specialized systems and a winnowing down of complexity has practical implications, coming into play in the design of a robot, especially in systems under development which place some of the process of design into the hands of an evolution-based processed. An article on such evolved machines describes this problem

> Where to start when designing the structure of a robot? Should it be randomly generated up to a certain complexity? Should it start from the simplest structures and grow in complexity, or should it start from the most complex designs and be simplified over the generations?[7]

This relationship of scales is a constant to evolution, not only the transitions such as that from the first single celled organisms to the dinosaurs and other behemoths which once walked the Earth back down to the small mammals from when we came, but also our own tenuous relationship to the microbiota we live in relative harmony with (only to have this or that tiny alien bit swoop in and defeat us...) Donna Haraway describes this well

> Some of these personal microscopic biota are dangerous to the me who is writing this sentence; they are held in check for now by the measures of the coordinated symphony of all the others, human cells and not, that make the conscious me possible.[8]

Bostrom's quote above concludes "The point is, we should not be confident that we have this under control". Likewise it would be shortsighted to assume our usurper – some manner

---

[7] Ibid.

[8] Haraway, Donna Jeanne. *When Species Meet*. Minneapolis: University of Minnesota Press, 2008. Pg. 4

of hyper complex and intelligent machine consciousness – will be any more in control of its eventual fate than we are. There is no final word in evolution

In a text about *The Invincible* by Wesley Osam (a sidebar describes these reflections as not "reviews or recommendations so much as attempts to work out my thoughts.") Osam decries the style which doesn't take its human characters as its main focus:

> But never mind the prose—Hard SF fans will tell you the ideas are the star! This argument has problems.
>
> First, if described badly enough even the most fascinating ideas can be boring. The Invincible's opening sets the tone. Before any of the characters even wake up it spends 500 words narrating a starship's automatic processes, and we're halfway through the first chapter before we get any dialogue that isn't tech jargon like "Full axis power. Static thrust." This novel cares more about things than people.[9]

This critique would likely not bother Lem – counterpoising a careful consideration of things to the normal focus on individuals, and examining the relationship between things and people is precisely the point he is driving at. The anthropocentric outlook, considering worthy of interrogation and focus only humans or things/beings closely resembling the human (either visually similar or being at least able to communicate with humans) is limiting; contemporary technological advances and philosophical and literary prodding are starting to test these limits. Machines are developed and or dreamed up which blur the distinction between conscious and automated, artificial and biotic. But as

---

[9] Osam, Wesley. "Stanislaw Lem, *The Invincible*." Recurring Bafflement. August 25, 2015.

importantly in Lem's writing, we are beginning to see how machines may evolve to a place where the drive for similarity imposed by humans begins to feel archaic, whether in the hyper-intellect of *Golem XIV*, the swarms of *The Invincible*, or the planet in *Solaris* – something altogether stranger.

Solaris

In Stanislaw Lem's _Solaris_, a planet-sized ocean/thing/being
– Lem once again blurs the distinction between these terms –
defies understanding. It is perhaps alive and conscious, but its
form is well beyond our working definition of sentience.
[fig. 39] Solaris exists in an orbit that defies normal physics,
accountable for only as the result of the actions of the ocean
whether because of chance or intention. Many years of
expeditions to Solaris (the planet where this being resides;
henceforth we'll use 'Solaris' to indicate this being and the
planet interchangeably) to investigate have come and gone.
The existence of Solaris has led to countless schools of
scientific, philosophical, and religious thought and as many
variations blending these three. But despite the observations
thus far made at much risk and cost of human life what has
been seen is at best contraindicating of any one way of
defining what Solaris is, what it represents.[1]

Solaris presents us with difference in its purest form:
impressive but mysterious, familiar yet uncanny. Lem's idea
in _The Invincible_ and _Solaris_ is a form of different creature as
dominant of its planet as we are but with no other familiar
aspects however close elements of each skew towards the
explainable. Solaris is a mind-body, "the dense plastic matter
of the planet directly embodies Mind."[2] It creates a number

---

[1] Compare this to the way in _The Invincible_ a group of scientists
from various fields encounter the alien swarm and one scientist
seems to correctly diagnose not only what it is, but also its entire
evolutionary history on the planet.

[2] Žižek, Slavoj, and John Milbank. _The Monstrosity of Christ:
Paradox or Dialectic?_ Cambridge, MA: MIT Press, 2009. Pg 91-92.
Žižek continues:

> This "spectral materialism" has three different forms: in the
> information revolution, matter is reduced to the medium of
> purely digitized information; in biogenetics, the biological
> body is reduced to the medium of the reproduction of the
> genetic code; in quantum physics, reality itself, the density

of complex, breathtaking phenomena: gigantic structures as much the flesh of the thing as some kind of 'weather', flaring up like huge fractal skyscrapers:

> When all's said and done, though, no terms can convey what goes on on Solaris. Its "dendromountains," its "extensors," "megamushrooms," "mimoids," "symmetriads' and "asymmetriads," its "vertebrids" and "rapidos"... [3]

And occasionally in the swirling formations there have been hints of the thing's capacity to read our thoughts, it forming objects drawn from explorer's memories. This later potential finds itself taken to an extreme in the events that transpire in the book. Our narrator is a scientist named Kelvin who arrives at the research settlement on Solaris to finds things very much awry. There are spectres haunting the station; each of the scientists aboard is dealing with a personal ghost, some character of significance from their past, which has suddenly sprung into being. Kelvin quickly acquires his own: a former lover Harey whose death he feels responsible for, and whose replicant immediately call forth many of the same feelings he felt for the original.

The story now tugs in at few different conceptual directions. The difficulty reconciling the doppelgängers' claim to consciousness and the relationship between our species and

---

of matter, is reduced to the collapse of the virtuality of wave oscillations (or, in the general theory of relativity, matter is reduced to an effect of space's curvature).
So late in the book, and I'm reminded of a term I had read a while back which is applicable to much of the whole of my project. Should I reach for this quote and its related thoughts later, will this footnote serve as a confession as to how late in the process I came upon the potent term 'Spectral Materialism'? Or should I not do so will this be the seed of work yet to come? Likely the latter.

[3] Lem, Stanisław. *Solaris*. San Diego: Harcourt Brace & Co, 2002.

the radically other are most intriguing for my purposes here. Lem's *Solaris* became films far more focused on a human-centric narrative than he intended. (Lem: "to my best knowledge, the book was not dedicated to erotic problems of people in outer space... This is why the book was entitled "Solaris" and not "Love in Outer Space"") This may result from two things, one, the general expectation (as expressed by the reviewer earlier about *The Invincible)* that narratives (whether film or book) focus on their humans and two, the way characters of *Solaris* find themselves, under very peculiar circumstances, dealing with very human issues of identity, recognition and guilt coupled with the sympathetic/elegiac manner in which Lem depicts these issues. The inner thoughts and desires of specific humans versus the species as a whole are given prominence in *Solaris* in a way they are not in *The Invincible* and *GOLEM XIV.* Žižek describes how the filmmaker Andrei Tarkovsky "does exactly the same as the lowest Hollywood producer, reinscribing the enigmatic encounter with Otherness into the framework of the production of the couple."[4] I'll avoid over analyzing the narrative and form of the novel and try to focus on the two themes I mentioned.

An oversimplification: Solaris shares with our liquid metal friend the ability to form any shape and to simulate the personalities of individual people. But what Solaris makes is durable, persistent versus the fleeting nature of the T-1000's impressions. And though Solaris's creations speak and are of a fleshy, familiar form, it is a mistake to assume the system or entity which created Harey and the others has done so in an attempt to commune with our species. Or that Solaris necessarily shares a grand agenda with Skynet or a short-term

---

[4] Žižek, Slavoj. *Less than Nothing: Hegel and the Shadow of Dialectical Materialism.* London: Verso, 2012. Pg. 658. This might be described in terms of the films focusing on one and another, with Lem more intrigued by the relationship between all(of us) and all else.

agenda of expedience with the T-1000 – why the strange duplicates come into being is never made clear; it is not something (necessarily) done for a purpose but rather something done and to be dealt with. The beings have no real conscious connection to Solaris; they do not know from whence they sprung. But they are clearly abnormal, for instance their attachment to the human from whose memories they were drawn is far more intense than maybe it was for their original/model, and should one try to kill one's spectre or place them in a rocket and blast them away from the planet, the next morning they will reappear, none the worse for wear. Except maybe they will be able to figure out what transpired, will be hurt by such actions – the copies are truly tragic figures, immortals not given a full bevy of lived history but rather drawn from the memory of someone close to them, both desired and unwanted by the soul from which they were conjured. Eric Wilson describes something similar in the form of the mummy:

> The mummy is a humanoid mechanism made of dead things but meant to prolong physical life indefinitely. This contradiction— death expanding life, life dependent upon death—cuts to the quick of the psychology of mummy making[5]

Solaris's copies begin with the same certainty of their consciousness that we have, the same certainty in their memories' validity. Eventually they may come to know they are copies and that their original is dead, simultaneously they realize their new lives are interminable. A version of Kelvin's dead wife is brought back not of her own volition; a parallel Harey is conjured which suffers the pain in forever being *and* forever being someone else. Harey grows despondent, suicidal, indirectly mirroring her original, she asks Kelvin (who she calls by his first name) "Tell me what I need to do so

[5] Wilson, Eric. The Melancholy Android: on the Psychology of Sacred Machines. 15-16.

I won't be there anymore. Kris. . ." To which he has no simple answer.

What if a lumpen piece of clay could indeed be animated? What if matter could be sufficiently given this or that identity, after which it would spring into consciousness, fully embodying the *self* imbedded within? It is not enough to discuss the likelihood of this coming to pass and/or what it would mean for us, so certain of our own immutable selves, so worried about what the ability to create consciousness or to resettle consciousness from one body to another will mean to our struggles, our ego. One must also consider what this will mean for the transferred consciousness, these reanimated, mummified versions of people that once were.

The threat is not in the ease of replication of consciousness somehow devaluing our own, but rather what comes later, once the deed is done. Kelvin phrases it well:

> Human beings set out to encounter other worlds, other civilizations, without having fully gotten to know their own hidden recesses, their blind alleys, well shafts, dark barricaded doors.[6]

With anything approaching AI and or/the transferal of a consciousness into a new body, all manner of foreseeable and unforeseeable moral and legal conundrums will arise, including right to self-regulation of one's existence, 'right to die' clauses, and worries about the way class will certainly determine access to digitally-assisted immortality are certainly relevant as well. But regardless of whether it is niche or universal, what it means for a mind to find itself sprung (back)into existence, how traumatic this might be and what types of heretofore unknown mental crises would arise is a huge unknown too often left out of the conversation. Full

---

[6] Lem, *Solaris*

disclosure: I won't do better than to begin here to think about this problem and slink off.

Solaris defies our definition of intention. As mentioned, the planet exists in an orbit that is impossible under normal circumstances – is this out of conscious decision or a survival adaptation more akin to a plant growing towards the sun? Is Solaris trying to commune with its visitors? When one squints in response to a bright light, it would be foolish for either the light or the person to assume the other was trying to speak. Likewise, Solaris's actions, whether obtuse in the form of structures appearing out of the ocean or oddly human-directed as in the re-creation of people from memories the scientists exploring it, can not be attributed to any particular intention and are as likely reflexes to specific stimuli as they are efforts to strike up a conversation.

There are a few basic directions humans tend towards when dealing with that which is alien: to anthropomorphize, to find a way to communicate with or barring this, to treat as enemy, to destroy. When a human receives an organ or limb transplant, the our body reacts not with a cool-headed analysis of that object's quality and use value but rather with immediate suspicion, quick to mark what is meant as a savior as an invader, something to be killed, rejected.

Cancer cells are the epitome of the familiar/unfamiliar. They are our cells *plus*, and its this simple plus which places the entire system at risk... they are of the same stuff but no longer acquiesce to the common goal of the organism, spreading and reproducing with no goal, no endgame, no chance of transmission like a normal pathogen such as bacteria, parasite or virus. Cancer cells despite their familiarity are that much more foreign than those three as they lack an ulterior motive, failing to perform their normal role while multiplying rapidly. They are a species unto their own, a short lived burst of innovation and life flaring up; if not completely excised they bite the hand that feeds them – they kill, extinguishing the only place they can survive.[7]

But in *Solaris* the defiantly other is not inside – however able to access what is inside our mind – but rather is something journeyed to: in this narrative we are what is invasive. Lacking a symbiot/host relationship, the desire to destroy Solaris or the swarm in *the Invincible* arises out of anything but self-defense:

> How foolhardy, how ludicrous this "victory at any price," this "heroic persistence of man," […] We were simply not cautious enough, we relied too much on our powerful weapons. We made mistakes, and now we must take the consequences. We and no one else are responsible.[8]

Golem XIV and Skynet are what we are semi-naïvely stumbling towards – something designed by humans, for humans, in the image of the human which may get out of hand, ignoring its maker's intentions and either harming or growing so indifferent to as to ignore humanity. But what if intelligences arise (or exist already) that are so defiantly different from our own as to resist either communication or the simple placing of their type of intellect at some place in a anthropocentric hierarchy.

Solaris presses against the delineations with which we define ourselves, all the more so as it presents some of what we consider fundamental features of our species as side effects, incidental. For instance, we think of empathy as key to us as

---

[7] There are a few examples of cancers being transmissible including Tasmanian devils and some species of clam, but generally speaking a cancer is a one-off occurrence. See: Shultz, David. "Contagious Cancer Found in Clams and Mussels." Science. June 22, 2016.

[8] Lem, *The Invincible*

social beings; Solaris shows itself as more than capable of seeing what the other is thinking, producing with an intense verisimilitude those we remember. But this kind of telepathy has the opposite effect of producing in us an understanding of the reader. From artist/author Jalal Toufic:

> The risk of a certain kind of telepathy is that it can extend the frame of what each person can perceive so far beyond its usual limits that the inconsistencies that are part of the world (or rather that are not part of the world, since they appear when one withdraws from the world or when the world withdraws from one) show up.[9]

Likewise, Solaris's 'reads' and subsequent creations render Solaris even less visible. They in fact go further, making both those into whose minds it peers somehow more opaque even to themselves, unable to control themselves – they know Solaris's creations are false and can't help falling in love again, they see the card up a sleeve and are nevertheless impressed by its emergence.

---

[9] Jalal Toufic, *Vampires*, 106

A conclusion of sorts (a brief reprise, a coda)

> The contemporary is he who firmly holds his gaze on his own time so as to perceive not its light but rather its darkness . . . [to] perceive, in the darkness of the present, this light that strives to reach us but cannot—this is what it means to be contemporary.
> *-Giorgio Agamben[10]*

> Just because we can do something, does it mean we should?
> "With all due respect, in this case I vote no."
> *-Shelly Fan, interviewing Dr. Natasha Kovacevic for an article about animal brains being wired together "into a functional organic computer", the "Brainet".[11]*

Science fiction is rarely simply 'art for art's sake' I'm reminded of this description of the way culture was encountered in the Middle Ages.

> When the medieval man looked at the column in the Souvigny abbey, with its four sides reproducing the wonderful ends of the earth through the images of the fabulous inhabitants of those regions—the goat-legged Satyr, the Sciapodes who moves on one foot, the horse-hoofed Hippopode, the Ethiopian, the manticore, and the unicorn—he had the aesthetic impression not that he was observing a work of art but rather that he was

---

[10] Giorgio Agamben, *What Is An Apparatus? and Other Essays*, trans. David Kishik and Stefan Pedatella (Stanford: Stanford University Press, 2009): 41-46

[11] Fan, Shelly. "Animal Brains Networked Into Organic Computer 'Brainet'" Singularity HUB. July 17, 2015.

measuring, more concretely for him, the borders of his world.[12]

Science fiction and its cast of strangers and monsters perform a similar task, mapping the state of things, and what's at stake. We eventually converge in time – if not in fact – with the future worlds predicted by fiction. Sometimes the predicted world resembles our own, as often it is more or less advanced than our own, making our circumstances seem inadequate (where's my jetpack?) or the predictions naïve such as the reliance on toggle switches and other manual inputs in space operas in which faster than light travel is a norm.

But lately, these two narratives – fiction and fact – have drifted closer and closer together. Ideas coming out of contemporary research institutions seem to leap beyond what is conjured in the imaginations of authors. Fictions that describe a near future such as the British anthology series *Black Mirror* and the drama *Humans,* and films like *Her* (Jonze, 2013) and *Ex Machina* (Garland, 2015) have become increasingly common.[13] Comic-based cinema feels the need to incorporate advanced but not uncommon materials like

---

[12] Agamben, Giorgio. *The Man without Content*. Stanford, Calif.: Stanford University Press, 1999. 34.

[13] Eerily similar, found six months after I wrote this bit I read a similar piece in *The Atlantic* which presents a reason for this convergence. From: Alsop, Elizabeth. "The Future Is Almost Now." The Atlantic. May 15, 2016.

> William Gibson sees as the end of speculation – the collapse of imagination into a reality that has already outpaced it? In other words, perhaps the reason writers and filmmakers are less inclined to imagine new "disasters" is that they're already adapting to so many. As Gibson explained in a 2007 interview, "I have to figure out what it means to try and write about the future at a time when we are all living in the shadow of at least a half a dozen wildly science-fiction scenarios.

carbon fiber into superhero costumes – Captain America's familiar bright tights replaced with a more muted, tactical armor.

Why do the authors of science fiction focus on lines of inquiry that so neatly dovetail with those of real-world theorists? Arthur C. Clarke put it well: "Science fiction is something that could happen - but you usually wouldn't want it to."[14] The gap between theoretical and applied science feels ever slighter as the newest ideas in science find their way into practical innovations and into our daily lives at an accelerated clip. Fields that seem on their face to be firmly beyond anything we would experience on a daily basis start to leak in to common experience. In quantum physics highly abstract and non-intuitive ideas are quickly becoming relevant: something as strange as quantum tunneling – whereby at very small quantum scales a particle may manage to pass through a solid barrier – starts to butt up against the miniaturization of computer chips which depend on the very certainty that the quantum world is so capable of undoing. Meanwhile a particle's ability to be neither here nor there (Schrodinger's Cat neither alive nor dead) becomes the basis for new types of cryptography, new types of computing etc. Science fiction is being redefined as "something *that is currently happening* though you *might not* want it to."

It has been great to work through the pile of science fact and fiction of which this publication is based. A great deal of it was released concurrently with my writing and continues to pour out as I try to finish. This is the end of the thing out of practicality, not due to either me running out of subjects to think about or things to say about them.

I find myself uncontrollably drawn to being either thankful or apologetic that you've read this thing. Thankful, in that you

---

[14] Clarke, Arthur C. *The Collected Stories of Arthur C. Clarke*. New York: Orb Books, 2000. Ix.

spent a finite amount of your time you could have (better?) spent otherwise, apologetic that it isn't somehow more conclusive, useful, or otherwise worthy of that time. I'll end a bit like I begun, avoiding the role of introduction/conclusion by talking about another ending, the final scene(s) of *Terminator 2: Judgement Day*.

Its always intriguing to find out parts one thought were most certainly fixed and essential to a film were actually in flux during production. Even though a liquid metal villain was already in Cameron's mind during the making of the first *Terminator* and was the leading candidate to play that part in the sequel, other ideas for a new antagonist were bandied about, including a version with two Arnolds (one good, one bad) and one with a female Terminator. The former ying-yang Arnold version seemed underwhelming as a sequel-worthy threat; the latter felt a bit ridiculous. But not so much as to remain permanently off the table as later films would go on to include a female liquid metal character. It is odd that gender could be considered relevant terms for such machines as female/male or even human/kangaroo are merely temporary states for a T-1000. Or perhaps what is suggested is that an initial gender identity bears definite implications both in the personality of the machine and in how we the viewers of the narrative will relate to the creature. Intriguing, but this is not where I'm going with this.

Rather I'll focus on something else that changed during production. There are two endings to *Terminator 2*. In the end of the theatrical release the T-1000 is defeated, thrown into a huge vat of molten metal. Presumably the T-1000 is diluted in the mass of liquid metal to an extent that it would be impossible for it to simple ooze and reform out of whatever the metal is when finally cast (an I-beam, a large public sculpture etc.; I find the image of the T-1000 oozing its way out of a Richard Serra sculpture very enticing) And the extreme temperature would irreparably damage its systems much as an earlier scene in which, partially frozen by liquid nitrogen, the T-1000's camouflage glitched/broke for a while.

The T-1000 gone, the arm from the first film is tossed into the molten metal. Schwarzenegger's good-guy Terminator then allows himself to be also melted down to protect against any chance of his technology inspiring would-be Skynet designers (as had happened with the remains of the Terminator in the original film). He stands on a platform that is lowered down into the metal by John Connor: "There is one more chip ... and it needs to be destroyed, also. *Here. (the Terminator points to his head)* I cannot self-terminate. You must lower me into the steel." This anti-suicide instinct makes very little sense – John Connor's command was sufficient to incite the T-800 to not kill anyone but apparently suicide is verboten. His calm decent into doom (complete with a wave goodbye) is ridiculous; it is hard to understand how the Terminator's AI doesn't extend this injunction against suicide into an accompanying sanction against *allowing* himself to die when easily avoidable. Nevertheless, we watch a robot's version of ashes to ashes, dust to dust, the robot being made formless, recycled. Just prior to completely being immersed a leather-clad glove reaches out and gives us a thumbs up. It is hard to watch this without stifling a laugh.

Regardless of this making (or not making) terribly much sense, we have in this ending a relatively tidy finish: not only has Skynet's plot to stop the future resistance leader from being born failed, the very triggers which would have resulted in Skynet's creation have been undone.[15]

---

[15] I know I know, causal loop, as having no Skynet built equals no Skynet to go back in time, fail, and trigger individuals to remove the possibility for it being built etc. Consider this yet another acknowledgement of the ease with which the film's logic breaks under gentle scrutiny.

But there was also a coda to the film scripted and filmed. In the cut final scene the factory sequence fades to white, after which we fade back into a 64-year-old Sarah Connor on a playground, where in voiceover she describes the future that didn't happen – 'Judgment Day' – as haunting her nevertheless. It is 2029; a glistening matte painting of a future skyline is in the distance, fashion has shifted somewhat as would be expected. John Connor is there, playing with what is presumably his child on a swing. [fig. 40] Sarah says John "fights the war differently than it was foretold. Here, on the battlefield of the Senate, the weapons are common sense... and hope."

The coda is a double-knot sealing the narrative, not impossible to untie should one desire to justify another film but a little more safe from accidental undoing. There is something unspoken in the way things turn out which echoes our earlier mentions of single-mindedness. In this tacked on future the threat to humanity posed by Skynet and its Terminators is/was the only one: once this single link is removed from the chain of things to come everything turns out great. Despite constant warnings to the contrary it turns out that war, climate change, overpopulation, and even malicious artificial intelligence fail to shake our progress towards a future much like the present, if a little glossier.

Cameron has expressed that his intention was for the ending of the theatrical version though lacking the expository element of the coda to imply a happily ever after in which the apocalypse never comes to be. He says he considered the two films a cohesive set with no need for sequels and in multiple interviews he has outlined his lack of interest in making more of the films. But this is hard to believe; examples of films that had wildly successful sequels that didn't lead to third and fourth films are few and far between. And I found while researching this text at least one interview in which a coy Cameron sidesteps news of a *Terminator 3*, giving the camera a wink and a nod.

And there's that arm left behind... in an earlier sequence in the factory the T-1000 manages to crush the good Terminator's arm. He rips free of his own arm and the film moves on leaving it behind - rife with the exact same future tech of the arm from the first film which inspired Skynet's designers which the film just made a point to show being destroyed. One can't help but feel should the following films have been a success (each of the three has done worse than the one previous) he would have gladly have claimed the lion's share of credit for the whole franchise, proudly describing how he laid the groundwork for what followed *Judgment Day*. This having not been the case despite continued efforts – three films and a television series – he can sit back and claim a distance from the terrible follow-ups. In order to retcon it helps to leave a trail of breadcrumbs heading down multiple paths.

Maybe what's left for me is to similarly lay out a path to either continue or call this finished in one fell swoop. About the latter, the best I can say is that my research is mostly finished. As I'm writing this it has thankfully been a slower week as far as pertinent text finding its way to my screen but I'd be fooling myself if I tried to pretend that another article pertinent to the many subjects I've covered wasn't right around the corner, ready to force my hand at adding another chapter or interlude between the now of this sentence and the now at the end of various stages of editing. Between *now* and the 'then-now' I will certainly give in to the temptation to write some more. But by the time these words get to the now of you reading this, that stack of uncommented upon experience, news, and new fictions will have doubtlessly grown but hopefully not to the point you regret spending time with this past me and my perspective delimited by my place in time.

I'm like the Terminator, slowly descending into the lava – I can pretend that calling this an end is not a choice, a suicide, but no one is fooled. Just like no one asked me to write this, no one has told me to stop writing so I've only myself to blame for its format, a book, failing the reality of its subject which flows into the future indifferent to maintaining a fealty to snapshots taken of it on the way. One might pose in a selfie knowing full well that one's outfit is not ideal, one's hair is about to get cut etc. and still take the time to smile.

Wilson's *The Melancholy Android* talks about robots in *Blade Runner* (Scott, 1982), for the most part overlooked in this volume (its so late in the game it seems foolish to bring up a story which could have easily been my foundation):

> [...] *Blade Runner* forms a definitive exploration of the psychological contradictions of golem making as well as a fascinating analysis of our postmodern propensity to loathe machines we love. Roy Batty, the film's golem, is designed to transcend matter but imprisoned in a material system. He oscillates between gnostic savior and gothic terror, Adam unfallen and Adam alienated.[16]

These poles could as easily be ascribed to the kinds of utopias and dystopias I've covered herein: technology as gnostic savior, and writers of fiction who see in the unfettered potentials of scientific advance the specters of future disaster.

In an article in *The Atlantic* by Ian P. Beacockon the history of our global system of timekeeping makes a good point:

> We ought to think more rigorously about who stands to gain from the smooth, efficient future we're being offered—and what we might lose.[17]

---

[16] Wilson, Eric. *The Melancholy Android: on the Psychology of Sacred Machines*. Albany: State University of New York Press, 2006. 88
[17] Beacock, Ian. "A Brief History of (Modern) Time." *The Atlantic*. December 22, 2015.

Much like the convenient occurrence of the singularity within its theorists' lifetime, the line between okay technological progress and bad is always drawn conveniently in the future and not in the past. Which is to say, things may be bad, but the *next* step will take us too far. This despite the possibility that the threshold beyond which dystopia has been irretrievably set in motion may sit forebodingly or more likely innocuously in our rear view mirror. What if doing our best to mitigate technology-driven change at this point only heightens the potential harm of those chain reactions initiated long before? What if what the technologists are selling as *disruptions* truly do hold a positive potential to abate the otherwise inevitable negative effects of past innovations, this or that next advance steering us clear of a wrong path we've already set out upon? What if neither a mitigation of technology's march nor allowing it to run headlong, full speed ahead will alter a bad course we are set upon; what if we are in *zugzwang* – a chess term for a position in which it is one's turn to move and every possible option makes things worse? What if our drive towards technology-fueled dominion over the earth was itself the gesture which made the end of it all less a matter of if than of when and/or how, or at the very least, capital as our opponent in a game in which any move against the position it has ensconced us in only tightens the rope?

Illustrations

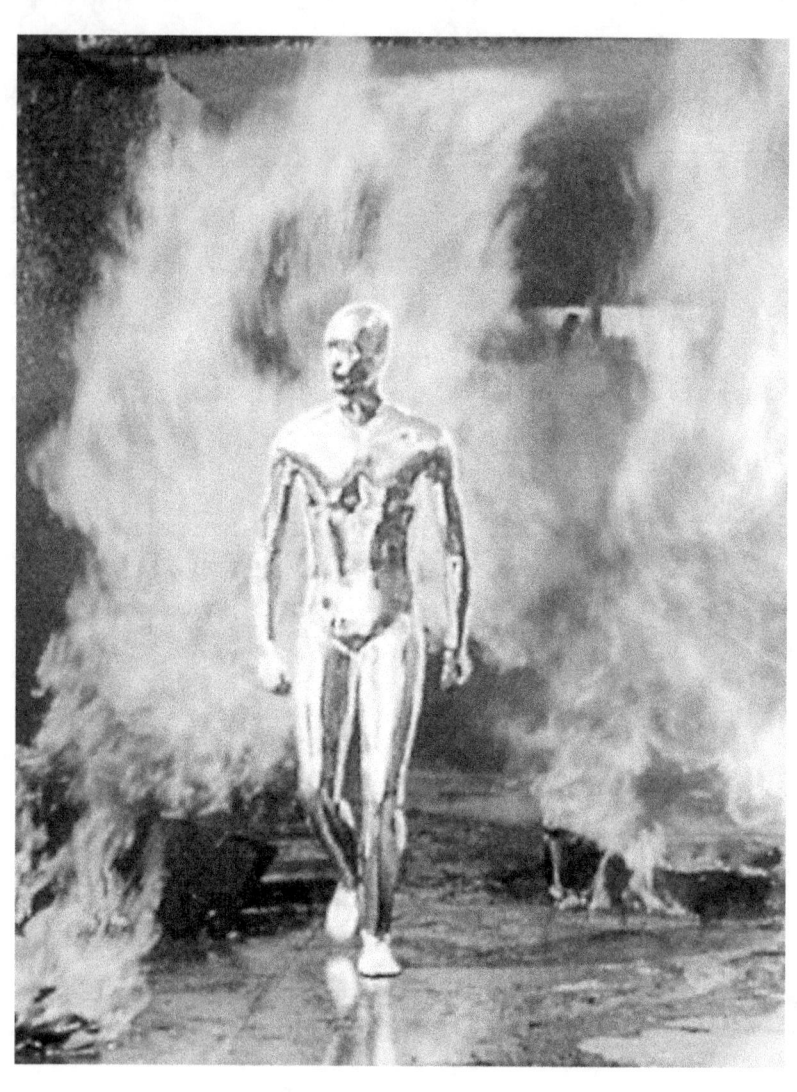

1. Still from *Terminator 2: Judgment Day* (Cameron, 1991)
   Pictured: T-1000

2. Still from *Der Golem* (Carl Boese, Paul Wegener 1920)

3. Illustration by Carlo Chiostri of Carlo Collodi's *Le Avventure di Pinocchio* (1901)

A TERRIFYING LOVE STORY.

A nthony Hopkins stars as a ventriloquist who uses the brash, abusive voice of his dummy to express his twisted desires to Ann-Margret. Based on William Goldman's best-selling novel, MAGIC is a steaming portrait of a love affair between a man who hides in the world of illusion and a beautiful woman desperately trying to recapture lost dreams. Together they are caught in a bizarre web of events that create pulse-stopping terror from the first frame of film!

1978, Color, 106 minutes, Thriller, **R**

Starring ANTHONY HOPKINS ANN-MARGRET
BURGESS MEREDITH and ED LAUTER
Produced by JOSEPH E. LEVINE
Directed by RICHARD ATTENBOROUGH
Screenplay by WILLIAM GOLDMAN

THRILLER

MAGIC

hi-fi
MONO
R
1501
C
VHS

4299-51501-3 1

NELSON
ENTERTAINMENT

MAGIC

Starring
ANTHONY HOPKINS · ANN-MARGRET
BURGESS MEREDITH · ED LAUTER

**THE LOVE THAT WILL HAUNT YOU...
FOREVER!**

4. VHS case from *Magic* (Attenborough, 1978)

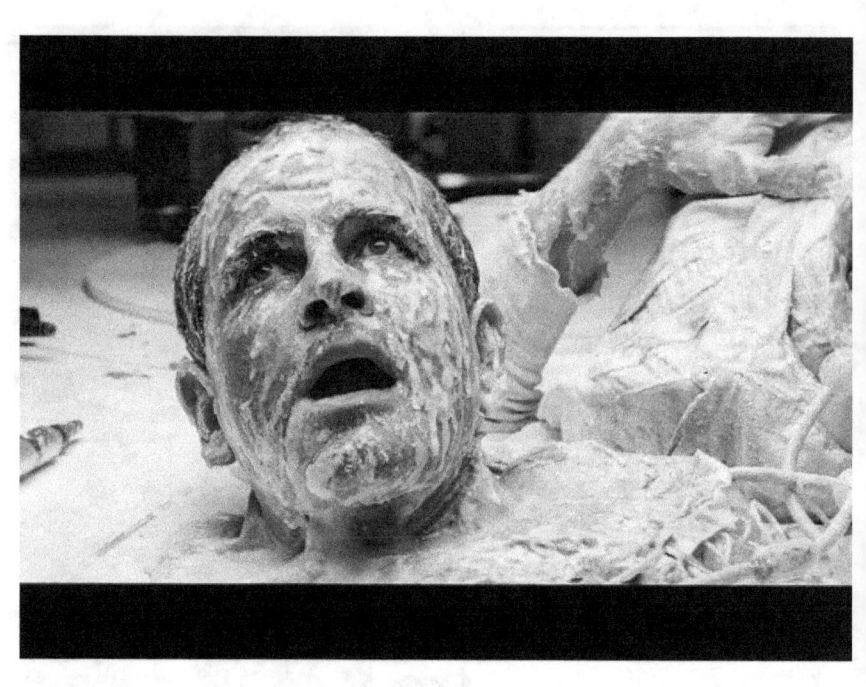

5. Ash from *Alien* (Scott, 1979)

6. Data (& Spot) from *Star Trek: The Next Generation*

7. (Left) Krusty the Clown and (Right) Homer Simpson

8. Poster from *Zelig* (Allen, 1983)

9. Image used for the poster of *The Thing* (Carpenter, 1982)

10. Still from *It Follows* (Mitchell, 2014)

ARPANET LOGICAL MAP, MARCH 1977

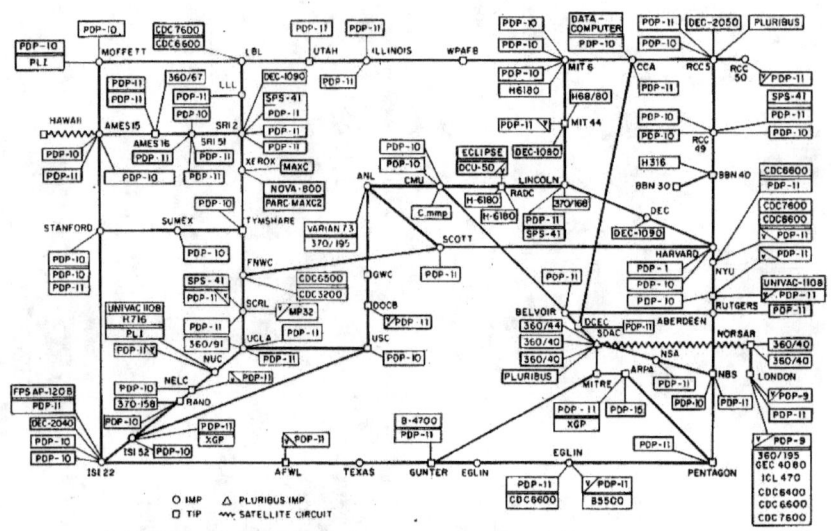

11. An early map of the proto-internet ARPANET

12. Poster from *Colossus: The Forbin Project* (Sargent, 1970)

13. Still from *2001: A Space Odyssey* (Kubrick, 1968)
Pictured: Dave and the insides of HAL-9000

14. Still from *Dr. Strangelove or: How I Learned to Stop Worrying and Love the Bomb* (Kubrick, 1964) Pictured: Dr. Strangelove

15. Cover of Peter Bryant's *Red Alert*

16. Seymour Skinner and Principle Seymour Skinner (aka Armin Tamzarian)

INVITATION
TO A
BEHEADING
VLADIMIR NABOKOV
A NOVEL
BY THE AUTHOR OF
LOLITA

17. The cover of Nabokov's *Invitation to a Beheading*

18. Still from *Adaptation* (Jonze, 2002)

19. The Noid

20. Still from *Day of the Dead* (Romero, 1985)
Pictured: Bub

21. Still from *World War Z* (Forster, 2013)

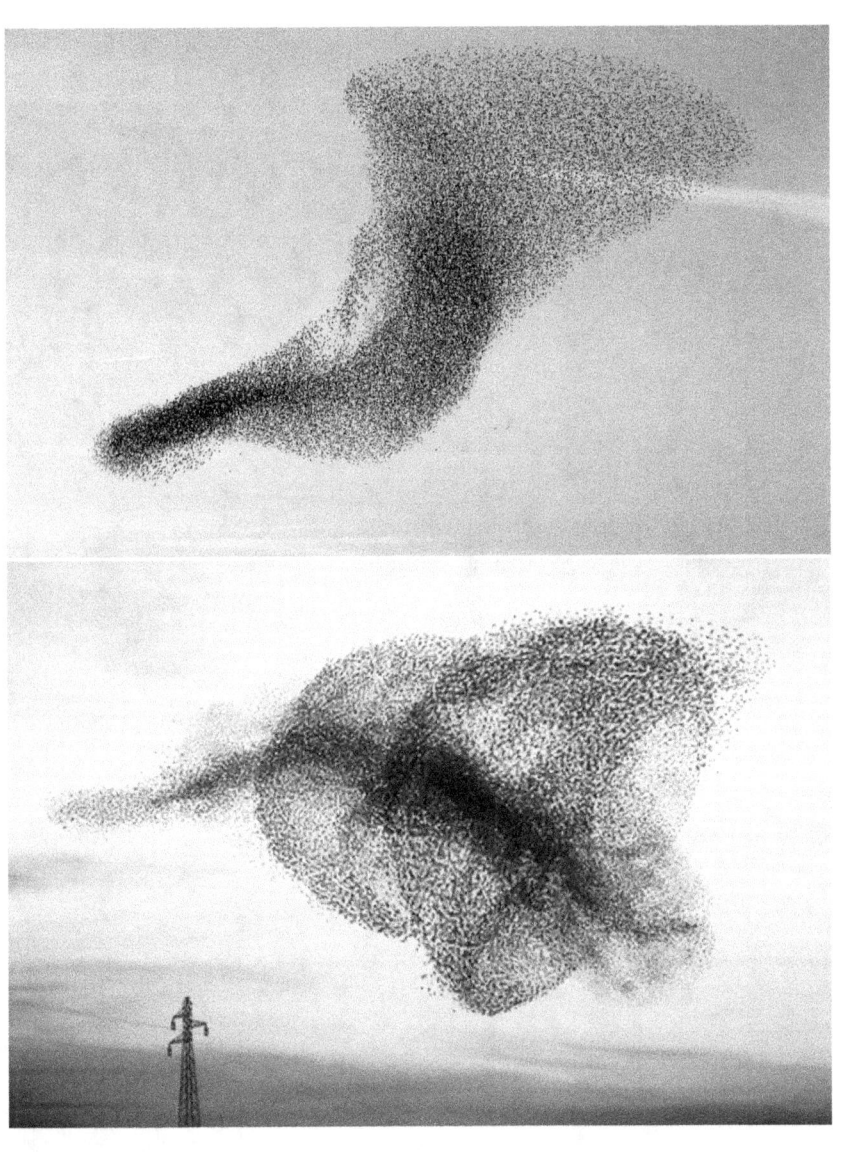

22. Bird swarms

23. Autonomous Flying Microrobots (RoboBees)

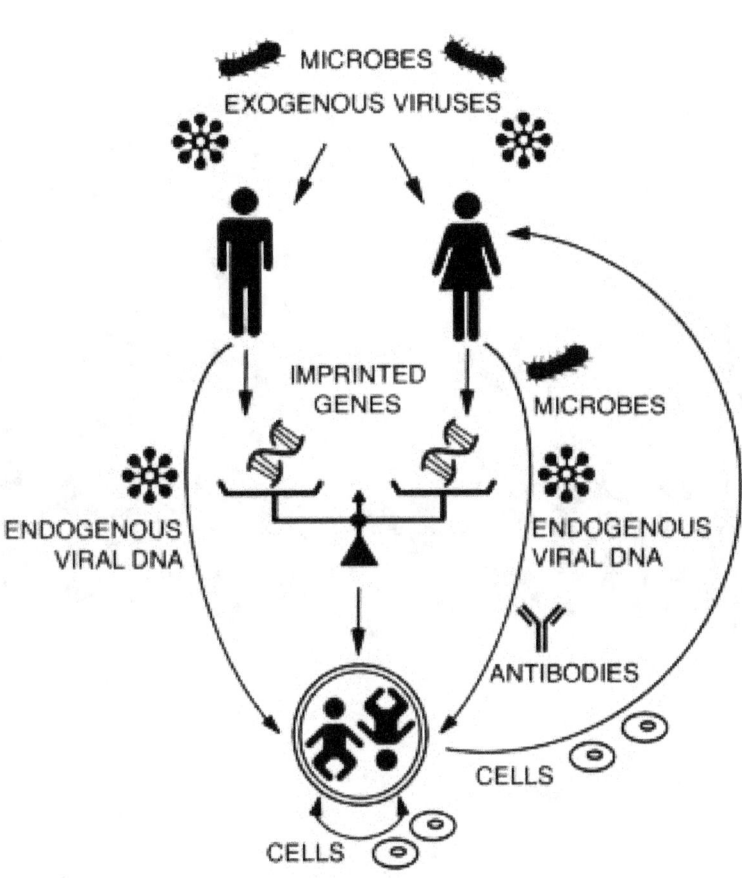

24. An illustration of 'human as superorganism'.

25. Still from *Lawnmower Man* (Leonard, 1992)

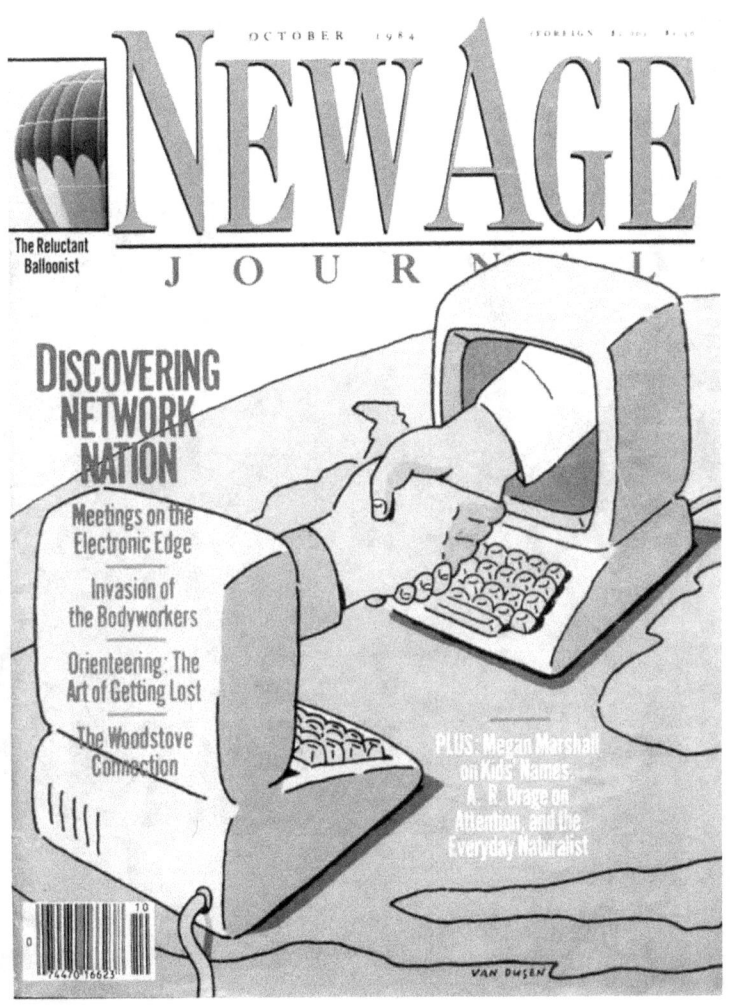

OCTOBER 1984 (FOREIGN

NEW AGE

JOURNAL

The Reluctant
Balloonist

DISCOVERING NETWORK NATION

Meetings on the
Electronic Edge

Invasion of
the Bodyworkers

Orienteering: The
Art of Getting Lost

The Woodstove
Connection

PLUS: Megan Marshall
on Kids' Names;
A. R. Orage on
Attention, and the
Everyday Naturalist

VAN DUSEN

26. *New Age Journal*, October 1984

27. Still from *Terminator 2: Judgment Day* (Cameron, 1991)
Pictured: John Connor & T-800 share a moment.

28. The Predator (image from Mortal Combat X, a fighting game featuring The Predator, Jason, Alien's Xenomorph among others)

29. Basil Rathbone portraying Sherlock Holmes

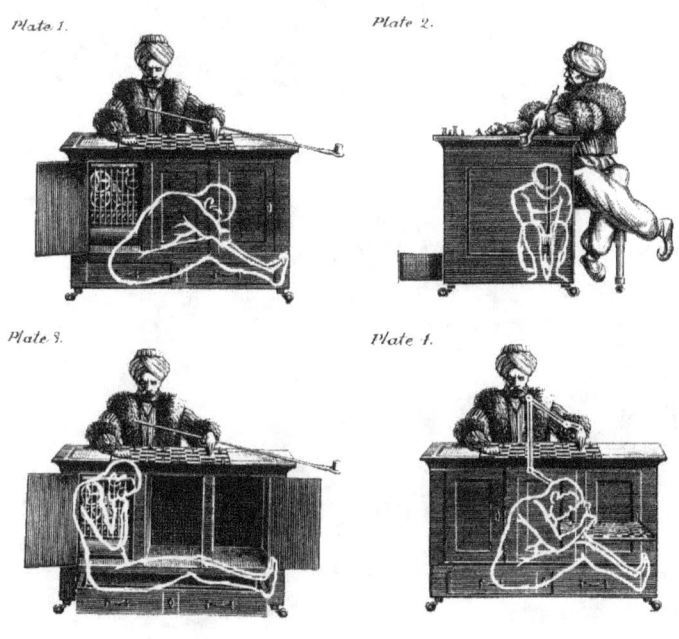

30. The Mechanical Turk

31. Gary Kasparov versus Deep Blue

32. Foster Partners model of an emergency services/supplies droneport

33. Illustration from Lewis Carrol's *Through the Looking Glass*, drawn by John Tenniel.

34. COTSbot

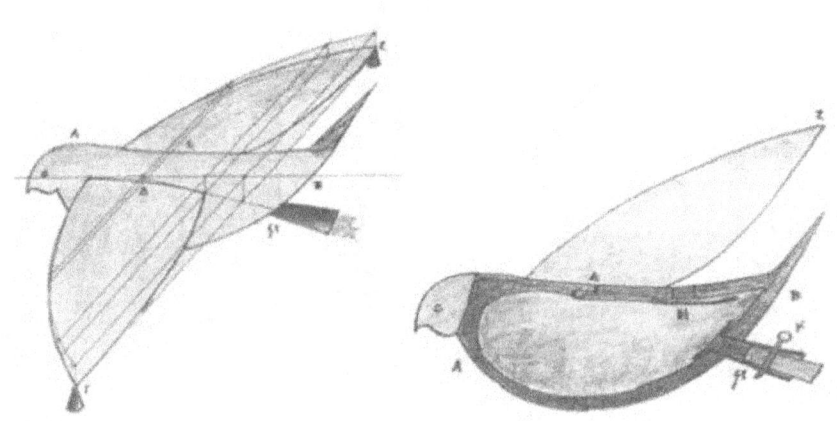

35. The steam-powered pigeon of Archytas

36. A chimera, attributed to Jacopo Ligozzi, Italian, 1547–1627

37. Still from *Willy Wonka & the Chocolate Factory* (Stuart, 1971)

ACE
SCIENCE
FICTION
SPECIAL
4

STANISLAW LEM
THE
INVINCIBLE

38. Cover of Stanislaw Lem's *The Invincible*

39. Japanese movie poster from *Solaris* (Tarkovsky, 1972)

40. Still from alternative ending to
Terminator 2: Judgment Day (Cameron, 1991)

Written, reasonably-well designed, and disastrously edited by Aaron Harbour.

Soft Machines was produced over the course of 2015-2016. Special thanks to The Luminary in St. Louis, Missouri where in early 2015 I was in residence and began getting serious about this becoming a full-fledged book-sized thing. And an extra special thanks to Jackie Im (and to our cat Casiopia) who allowed me countless nights to drink beer and work on this project. Without her presence, patience, inspiration and constant challenge I'd never be able to finish such an undertaking.

-

A note on how this book was (self) published. I have made many books and other publications at a local copy shop and/or by hand; in this instance I wanted a book-like book. Ideally it would be perfect bound with lots and lots of pages. I found an online publishing site called Create Space, a subsidiary of Amazon. The process of working with Create Space aligned strangely with some of the concerned expressed herein. I uploaded the cover and contents online of an initial proof version just to see if the thing would show up at my door looking like a nice book. What followed was a series of required adjustments to my file: I would submit PDFs and in 12-24 hours I'd get an email in response outlining what needed to change. This slow process repeated until after about a week everything was in order. Was an algorithm checking my files or was a human opening them each time? If a human, were they at a job site somewhere or was it a kind of distributed, cloud-based Amazon Mechanical Turk type process? If it was an algorithm, did it recognize itself-as-subject?

Aaron Harbour does a few too many things. He is an artist, curator, writer and DJ operating out of Oakland, CA. As an artist, his work concerns identifying 'misbehaving objects, words and things acting against expectation with which he attempts to collaborate. He has shown work and or performed at The Luminary, St. Louis; Windo Space, Los Angeles; City Limits, San Francisco; Southern Exposure, San Francisco; Gaylord's, Oakland; Asian Art Museum, San Francisco; New Langton Arts, San Francisco.

He is co-director of Et al. & Et al. etc, a gallery program in San Francisco, and has additionally curated exhibitions at The Popular Workshop, Important Projects, MacArthur B Arthur, Interface Gallery, Liminal Space, the Royal NoneSuch Gallery, and various art fairs.

He runs *Curiously Direct*, an art criticism website and blog at curiouslydirect.com and on Facebook, and has additionally written for *Fillip Magazine*, *San Francisco Arts Quarterly*, *Art Practical*, *Decoy Magazine*, *Art Cards*, and several small publications/artist catalogues.

He has DJed widely in the Bay Area; he has produced 100+ episodes of his ongoing podcast series is called *Timber*, full show archive is at kgpc969.org/timber. He received some modicum of education at the San Francisco Art Institute.